ANNALES ACADEMIÆ SCIENTIARUM FENNICÆ

MATHEMATICA

DISSERTATIONES

154

AF855074

TOPOLOGICAL STABILITY THROUGH TAME RETRACTIONS

AASA FERAGEN

University of Helsinki, Department of Mathematics and Statistics

To be presented, with the permission of the Faculty of Science of the University of Helsinki, for public criticism in Auditorium XII, the Main Building of the University, on December 14th, 2009, at 12 o'clock noon.

HELSINKI 2009
SUOMALAINEN TIEDEAKATEMIA

Copyright © 2009 by
Academia Scientiarum Fennica
ISSN-L 1239-6303
ISSN 1239-6303 (Print)
ISSN 1798-2375 (Online)
ISBN 978-951-41-1036-8 (Print)
ISBN 978-951-41-1045-0 (PDF)

Received 12 November 2009

2000 Mathematics Subject Classification:
Primary 57R45; Secondary 58K60, 58K25, 58K70.

YLIOPISTOPAINO
HELSINKI 2009

To my grandmothers
Magnhild and Agnes
who showed me how to be strong
through will
and love of life

Acknowledgements

Every student is indebted to her advisor, and I have had the great fortune of having two.

I thank Andrew du Plessis, my mentor for the past three years, for teaching me, challenging me and supporting me – all when I needed it the most. His scientific input has taught me to see problems from angles I thought were impossible – at least for me. Through my work with him I have been subjected not only to topology, but to its interactions with algebra, algebraic geometry and differential geometry, and I owe him my deepest thanks for giving me such a strong and diverse geometry toolbox.

I thank Sören Illman for teaching me transformation groups, which I find gives a beautiful way to view even other areas of mathematics. I am also thankful to him for giving me support and freedom when I chose to change my field.

I want to direct my warmest thanks to Erik Elfving for introducing me to topology, and to Kalevi Suominen, whose lectures were simply the best.

This is the moment, if ever, to look back to my teachers. I thank Tor Westberg for proving that math was my thing, although I did not particularly want to see it at the time. At one point I claimed that I would never study mathematics again – and I have never been more wrong! I thank Hans Olav Sunsby for changing my mind.

I thank the Department of Mathematical Sciences at Aarhus University for their hospitality during my time as a guest student there.

I wish to thank the staff at the Department of Mathematical Sciences at Aarhus University and at the Department of Mathematics and Statistics at the University of Helsinki for their kind and efficient assistance. I especially want to thank Lars Madsen and Martti Nikunen for their technical help throughout the years, and Mika Koskenoja and Mats Gyllenberg for their efficient help in the months leading up to the defence.

I thank the pre-examiners Professor Hans Brodersen and Professor Maria Aparecida Soares Ruas, whose careful reading and valuable comments helped me improve the quality of the final result.

For financial support I thank the Magnus Ehrnrooth Foundation, the Helsinki University Science Foundation and the Topological Transformation Groups Research Group at the Department of Mathematics and Statistics, University of Helsinki. I also thank the Topology Group at Aarhus University as well as the Oskar Öflund Foundation for travel support.

Like any PhD student, I would not have kept my sanity, had it not been for my friends and colleagues, who have always had time for mutual therapy – often aided by an arsenal of unhealthy beverages. To all my friends, officemates and colleagues: you meant a lot!

I especially want to thank my colleagues Alex Nuija, who made me realize that I was good enough; Cristina Martinez and Guillaume Theret, who kept telling me, day after day, that at some point I would be less confused; and Khalid Rian for his good company and for knowing exactly when to talk – and when not to.

I thank my friends:

Samantha Herbst – my parallel translate since our first week of university, when we measured the volume of the IU chemistry building together in terms of short and long sticks;

Ansku&Michael Patzschke, with whom I always had the most interesting conversations, and Anna Pomoell, with whom I always had the wildest ideas;

Muwan Chen, Fabia Febbaro, Cristina Esteva Font and Baba Muhammad (they were there from the start) for being my family away from home, together with Gülnur Dogan and Çanan Doganli (my never failing suppliers of astrophysics espresso coffee, turkish meatballs and warm hugs), Alessandro Cuoco (my partner-in-crime for cakes and scary movies) and Jyoti Rajput (always there with calm advice and tea).

Some people never get the opportunity to stick their head in the sand and pretend they do not hear you. I want to thank my family – my sister Maren and my parents Eva and Harald – for their patience and support.

Last, and most, of all, I thank Jerker, for being my best friend, adviser, slavedriver, housewife, computer doctor, and love.

Aarhus, November 2009

Aasa Feragen

Contents

Introduction

A smooth map is said to be stable if small perturbations of the map only differ from the original map by a smooth change of coordinates. This means, in particular, that small perturbations of the map do not change the differential-geometric properties of the singularities of the map.

The notion of a stable map was invented in order to classify mappings in terms of their singularities, and the roots of the theory go more than 50 years back. The oldest stability theorem in the book is perhaps the Morse lemma, which completely characterizes the singularities of a stable function (or, as it is called in Morse theory, a Morse function). However, the real foundations of the theory of stable maps were laid by people like Hassler Whitney, René Thom and John Mather.

The definition of a stable map was formulated by Whitney in the 1940s. Thom, through the development of his Catastrophe theory in the 1960s, classified the singularities and stable deformations of generic smooth maps in low codimensions. This work did not only form the basis for more general developments, but has also been used extensively in practical applications where it is useful to recognize maps through the development of their singularities in deformations, for instance in image analysis.

Through a famous series of papers written in the 1960s and 70s, John Mather worked out the general theory of stable maps, founding what is often referred to as *Mather theory* [Mat68b, Mat69b, Mat68a, Mat69a, Mat70b, Mat71]. He gave a number of different definitions of smooth stability, and through them a number of different tools to attack stability problems using as different fields of mathematics as analysis, topology and commutative algebra. He also showed that among proper maps, the stable maps are generic exactly when the source and target dimensions of the maps belong to the so-called *nice dimensions*, which in particular means that the applicability of the theory of stable maps is limited to those dimensions.

This restriction led to the formulation of *topological* stability, where the smooth coordinate changes are replaced with homeomorphisms. Many of the different definitions of smooth stability have natural topological counterparts, but these do not typically give equivalent definitions. Mather also studied topological stability, and among his most prominant results are the result that topologically stable maps are generic among proper maps for all source and target dimensions, and Mather's *transversality theorem*, where he defines a canonical stratification of jet space with the property that topologically stable maps are characterized by their transversality to the stratification [Mat70a]. An alternative account can be found in [GWdPL76]. Unfortunately, Mather's stratification is nearly impossible to use in practise, as the stratification is very hard to compute.

In their book [dPW95], Andrew du Plessis and Terry Wall present the state-of-the-art theory of topologically stable maps, and in particular they give necessary and sufficient conditions for topological stability. Their results include a transversality theorem which gives sufficient conditions for topological stability through transversality to a stratification of jet space by so-called *civilized* submanifolds. These civilized submanifolds are closely connected with the tame retractions which are the subject of this thesis.

Stable maps play the main characters in several active branches of research. There is ongoing work on topological stability, where many open questions remain [dPW95, dPW04]. Related to this, there are recent results on topological and bi-Lipschitz equivalence and triviality for smooth singularities [Dam88, Dam92, DG93, FR04, BCFR07], as well as on topological sufficiency of jets [Sku09]. Related to the E-tame retractions studied in this thesis are also the (a)-regular foliations used by Murolo and Trotman [MT06] to study a smooth version of the Whitney fibration conjecture. Even smooth stability is still a research topic; for instance there has been much recent research concerning invariants of smoothly stable maps [HMdJRF07, ORF09, ORF09'], and ideas from the theory of stable maps is also applied to other sciences such as obot kinematics or image analysis [Dam95, DG99, Dam99, ON06].

The problem at hand. In this thesis we study the following problem related to topological stability of maps:

Let $f\colon \mathbb{R}^n \to \mathbb{R}^p$ be a weighted homogeneous map or germ with a ministable, weighted homogeneous unfolding $F\colon \mathbb{R}^n \times \mathbb{R}^d \to \mathbb{R}^p \times \mathbb{R}^d$ of the form $F(x,\underline{u}) = (\tilde{f}(x,\underline{u}),\underline{u})$ for a deformation $\tilde{f}$.

If we omit some of the unfolding variables u_i (where $\underline{u} = (u_1,\ldots,u_d)$), the resulting unfolding $\tilde{F}\colon \mathbb{R}^n \times \mathbb{R}^c \to \mathbb{R}^p \times \mathbb{R}^c$ is no longer stable, since F was ministable. Is it possible that $\tilde{F}$ is still topologically stable? How many unfolding variables can we remove without losing topological stability?

We study this problem when f is a germ belonging to the series $E_{p,0}(*)$ and $Z_{p,0}(*)$ with $\mathscr{K}$-normal forms

$$\begin{cases} y^3 + yx^{2p}W_{p-1}(x) + x^{3p}, \\ x(y^3 + yx^{2p}W_p(x) + x^{3p}), \end{cases}$$

where $W_p(x) = w_0 + w_1x + w_2x^2 + \ldots + w_{p-1}x^{p-1}$, $w_0 \neq 0$.

Our hypothesis is that the unfolding

$$F^+\colon \mathbb{R}^n \times \mathbb{R}^c \to \mathbb{R}^p \times \mathbb{R}^c$$

obtained by omitting the non-positively weighted unfolding variables from the standard ministable unfolding F of f, is a topologically ministable unfolding of f. We verify this for $p \leq 4$ in the E-series, and for $p \leq 3$ in the Z-series, and discuss what is still needed for the general case.

The idea of the proof is to construct an *E-tame retraction* $F \to F^+$. E-tame retractions form a class of retractions satisfying certain geometric conditions, whose main property is that they preserve topological stability from smoothly stable unfoldings.

The thesis is organized as follows: In Chapter 1 we give basic definitions and state results that will be used later in the thesis. We briefly review the necessary theory of smooth maps and their stability (or lack of it). We also fix the terminology which will be used throughout the paper.

In Chapter 2 we define E-tame retractions, and discuss the relation between E-tame retractions and topological stability of unfoldings. Moreover, we present basic properties of E-tame retractions and prove some results on how to construct them, which are important tools.

In Chapter 3 we study the group $\mathscr{A}_f$ of smooth coordinate changes leaving a multigerm f fixed. We prove that $\mathscr{A}_f$ has a maximal compact subgroup $MC(\mathscr{A}_f)$, unique up to conjugation, and that the quotient $\mathscr{A}_f/MC(\mathscr{A}_f)$ is contractible. Moreover, we show that $MC(\mathscr{A}_f)$ can be decomposed as a product $\prod MC(\mathscr{A}_{g_i})$, where the g_i are monogerms representing the components of f. These results, which form a nice theorem on their own, will be useful in solving local-to-global problems in constructions of E-tame retractions.

At the end of the chapter we prove some results concerning how large the maximal compact subgroup can be, and compute the maximal compact subgroup for (multi)germs whose components belong to the germ classes $E_{p,0}(*)$ or $Z_{p,0}(*)$.

Chapter 4 is dedicated to constructing E-tame retractions $F \to F^+$, which is the main purpose of the thesis. We go through the construction in great detail for the E-series singularities, and note that the same type of construction will work also for the Z-singularities. The construction for the $E_{p,0}(*)$ singularities with $p \leq 4$ is done in several steps:

- Parametrize the instability locus of the positively weighted unfolding F^+, which is where the construction of local E-tame retractions is the most difficult. Show that the positive instability locus is, in fact, a stratified set with respect to presented singularity types.
- See that the singularity types occurring on the positive instability locus outside the non-positively weighted subspace, are actually $E_{p,0}(*)$-singularities for lower p. This suggests an inductive procedure, as we are now able to construct local E-tame retractions outside the non-positively weighted subspaces.
- Solve the resulting local-to-global problem on level sets of a distance function measuring the distance from the non-positively weighted subspace, by

controlling the geometry of the stratified set comprising the instability locus. In practise, this involves forcing E-tame retractions to coincide as we pass to lower-dimensional strata.

- Extend the level set retraction to a neighborhood of the non-positively weighted axis using the $\mathbb{R}^+$-action associated to the weighted homogeneity.

We discuss the similarities and differences between the E-series and the Z-series cases, and sketch the construction of an E-tame retraction $F \to F^+$ for the singularities $Z_{p,0}(*)$ with $p \leq 3$, paying extra attention to the differences between the cases.

Finally we discuss the problems connected with the construction for higher p.

1. Preliminaries

1.1. **Stable maps and their friends.**

In this section we give a short sketch of the parts of the classical singularity theory of smooth maps which we will need for our constructions.

We would like to emphasize the following notation, which will be used extensively throughout the thesis: Given a map $f\colon N \to P$, we will denote by $t(f)$ the target P and by $s(f)$ the source N.

We also agree that whenever we talk about *stable* maps or germs, we mean *smoothly stable* maps or germs, as defined below. Whenever we have some other form of stability in mind, this will be stated explicitly.

For further questions related to terminology and notation, we refer the reader to the Index at the end of the thesis.

1.1.1. *Stable maps.* A C^∞-smooth map $N \to P$ is said to be *smoothly stable* if there exists a neighborhood $\mathscr{U}$ of f in the strong topology (also called the Whitney topology) on $C^\infty(N, P)$ such that for any $g \in \mathscr{U}$ we can find diffeomorphisms Ψ and Φ making the diagram

$$\begin{array}{ccc} N & \xrightarrow{\ f\ } & P \\ {\scriptstyle\Psi}\downarrow & & \downarrow{\scriptstyle\Phi} \\ N & \xrightarrow[\ g\ ]{} & P \end{array}$$

commute. If f is a proper map, then the diffeomorphisms $\Psi \in \mathrm{Diff}^\infty(N)$ and $\Phi \in \mathrm{Diff}^\infty(P)$ can be chosen to depend continuously on $g \in C^\infty(N, P)$, when all mapping spaces are given the strong topology [Mat69b, Theorem 3.2].

The group $\mathscr{A} := \mathrm{Diff}^\infty(N) \times \mathrm{Diff}^\infty(P)$ acts on $C^\infty(N, P)$ through the formula $(\Psi, \Phi) \cdot f = \Phi^{-1} \circ f \circ \Psi$, and an equivalent statement of the definition of a stable map is that f is stable if its $\mathscr{A}$-orbit is open in $C^\infty(N, P)$.

There is an alternative notion of stability, defined by Mather – the so called *infinitesimal stability*.

Denote by θ_f the set of maps $\theta\colon N \to TP$ such that the following diagram commutes:

$$\begin{array}{ccc} & & TP \\ & \overset{\theta}{\nearrow} & \big\downarrow{\scriptstyle \pi_P} \\ N & \xrightarrow[f]{} & P \end{array}$$

The maps $\theta \in \theta_f$ are called *vector fields along* f and we denote $\theta_N = \theta_{\mathrm{id}_N}$, $\theta_P = \theta_{\mathrm{id}_P}$. Define maps $tf\colon \theta_N \to \theta_f$ and $wf\colon \theta_P \to \theta_f$ by setting

$$\begin{array}{l} tf(\xi) = Tf \circ \xi, \\ wf(\eta) = \eta \circ f. \end{array} \qquad \begin{array}{ccc} TN & \xrightarrow{Tf} & TP \\ {\scriptstyle \xi}\big\uparrow & \overset{\theta}{\nearrow} & \big\uparrow{\scriptstyle \eta} \\ N & \xrightarrow[f]{} & P \end{array}$$

We say that f is *infinitesimally stable* if $tf(\theta_N) + wf(\theta_P) = \theta_f$.

Theorem 1. [Mat69b] *A proper smooth map is stable if and only if it is infinitesimally stable.* □

1.1.2. *Stable germs.* Consider the mapping space $C^\infty(N, P)$ and fix a subset $S \subset N$. We define an equivalence $\sim$ on $C^\infty(N, P)$ by saying $f \sim g$ if there exists a neighborhood U of S in N such that $f|U = g|U$. We denote the equivalence class of f by $\hat{f}\colon (N, S) \to (P, f(S))$, and call it the *germ of f at S*. If we want to emphasize the set S, we may also use the notation $\hat{f}_S$. We say, furthermore, that the maps f and g are *representatives* of the germ $\hat{f}$.

One can also define germs of maps which are not smooth. If we wish to emphasize that we are considering germs of smooth maps, we say "smooth germ".

In some cases, a germ does not come from an obvious representative, and in these cases we may choose to denote our germ by $f\colon (N, S) \to (P, y)$ rather than $\hat{f}$.

We will mostly consider situations where S is a finite set and $f(S)$ consists of a single point y. Whenever S consists of a single point x, we call $f\colon (N, x) \to (P, y)$ a *monogerm*, and when S consists of several points, we call $f\colon (N, S) \to (P, y)$ a *multigerm*.

Note that the properties of the germ f only depend on what happens in arbitrarily small neighborhoods of S and y; hence by passing to coordinate charts, it suffices to consider monogerms

$$(\mathbb{R}^n, 0) \to (\mathbb{R}^p, 0)$$

and multigerms

$$\bigsqcup_{i=1}^{s} (\mathbb{R}^n, 0) \to (\mathbb{R}^p, 0),$$

and we shall pass freely between the coordinate chart notation and the "global looking" version as it suits our purposes.

We denote by $\mathscr{E}(n)$ the ring of function germs $(\mathbb{R}^n, 0) \to \mathbb{R}$, and denote by $m(n)$ the ideal in $\mathscr{E}(n)$ consisting of germs that map 0 to 0. We write $\mathscr{E}(n,p)$ for the set of germs $(\mathbb{R}^n, 0) \to \mathbb{R}^p$, and note that $\mathscr{E}(n,p) \cong \bigoplus_p \mathscr{E}(n)$. If the dimension n is obvious, and we rather want to emphasize the point $y \in P$ at which we are taking our germs, then we write m_y instad of $m(n)$.

We proceed to define stability for germs. Note that the original definition does not carry over directly, since it does not make sense to define a strong topology on the space of germs $(\mathbb{R}^n, 0) \to (\mathbb{R}^p, 0)$ – any germ $(N, S) \to (P, y)$ has representatives on arbitrarily small neighborhoods of S which are far from each other in the strong topology. It is possible to get around this problem by defining stability via representatives [dPW95, Chapter 4.4], but it is easier and more customary to define stability of germs through infinitesimal stability.

Let $f\colon (N,S) \to (P,y)$ be a smooth germ. Define the set θ_f of *vector fields along* f to be the set

$$\{\theta\colon N \to TP | \pi_P \circ \theta = f\},$$

and set

$$\begin{aligned} \theta_S &= \theta_{(N,S)} = \theta_{\mathrm{id}_{(N,S)}}, \text{ and} \\ \theta_y &= \theta_{(P,y)} = \theta_{\mathrm{id}_{(P,y)}}, \end{aligned}$$

where we choose notation depending on how explicit we need to be. We define $tf\colon \theta_S \to \theta_f$ and $wf\colon \theta_y \to \theta_f$ by setting

$$\begin{aligned} tf(\xi) &= Tf \circ \xi, \\ wf(\eta) &= \eta \circ f, \end{aligned}$$

and define f to be infinitesimally stable if $\theta_f = wf(\theta_y) + tf(\theta_S)$.

Moreover, we agree to call germs *stable* whenever they are infinitesimally stable.

One would expect some sort of correspondence between local and global stability, and using the definition of stability for germs, we are able to formulate this through a third definition of stability. We say that a map $f\colon N \to P$ is *locally stable* if the germ $\hat{f}\colon (N, \Sigma f \cap f^{-1}(y)) \to (P, y)$ is stable for each $y \in \mathrm{Im}(\hat{f})$. We shall see that stability is equivalent to local stability for most maps.

We say that a stable germ $f\colon \bigsqcup_s (\mathbb{R}^n, 0) \to (\mathbb{R}^p, 0)$ is *ministable* if it is not $\mathscr{A}$-equivalent to any germ $\tilde{f} \times \mathrm{id}_{\mathbb{R}^d,0}$, where $d > 0$ and $\tilde{f}\colon \bigsqcup_s \mathbb{R}^{n-d} \to \mathbb{R}^{p-d}$ is stable.

1.1.3. *Unfoldings.* One of the ways to handle non-stable maps is to "embed" them in maps which *are* stable, more precisely in *unfoldings*. In fact, having a stable unfolding is a generic property among proper maps. We shall get back to this shortly, but first, we need some definitions.

Given a smooth map $f\colon N \to P$, an *unfolding* of f is a triple $\{F; i, j\}$ where F is a smooth map and i, j are embeddings, such that the diagram

$$\begin{array}{ccc} N & \xrightarrow{\ f\ } & P \\ {\scriptstyle i}\big\downarrow & & \big\downarrow{\scriptstyle j} \\ N' & \xrightarrow[\ F\ ]{} & P' \end{array} \tag{2}$$

commutes and is Cartesian, and where $j \pitchfork F$.

We can often construct a less general – and more practical – notion of unfolding; namely the parametrized unfolding. A parametrized unfolding is an unfolding as in (2), where $N' = N \times U$ and $P' = P \times U$ for some neighborhood U of 0 in $\mathbb{R}^d$, $F = (F_P, \mathrm{pr}_U)$ and i, j are the 0-level diffeomorphisms of N and P onto $N \times \{0\} \subset N \times U$ and $P \times \{0\} \subset P \times U$. In particular, since $F(x, u) = (F_P(x, u), u)$, we say that F is U-level-preserving, and write $F \in C^\infty_{\mathrm{lp}}(N \times U, P \times U)$.

Whether or not an unfolding can be considered (up to diffeomorphisms of source and target) as a parametrized unfolding, depends on whether or not the source- and target subspaces admit trivial tubular neighborhoods.

Both these definitions of "unfolding" extend naturally to definitions of unfoldings for germs, and in fact, for germs at points, we can always turn unfoldings into parametrized unfoldings by choosing suitable local coordinates.

Given two unfoldings $\{F_1, i_1, j_1\}$ and $\{F_2, i_2, j_2\}$ of a smooth map f we define a morphism $F_1 \to F_2$ to be a pair of smooth maps (Φ, Ψ) such that the following diagram commutes:

$$\begin{array}{ccccccc} N_1' & & & \xrightarrow{\qquad F_1 \qquad} & & & P_1' \\ & \nwarrow^{i_1} & & & & \nearrow^{j_1} & \\ {\scriptstyle \Psi}\big\downarrow & & N & \xrightarrow{\ f\ } & P & & \big\downarrow{\scriptstyle \Phi} \\ & \swarrow_{i_2} & & & & \searrow_{j_2} & \\ N_2' & & & \xrightarrow[\qquad F_2 \qquad]{} & & & P_2' \end{array}$$

We say that an unfolding $\{F_2, i_2, j_2\}$ of f is *versal* if, for every other unfolding $\{F_1, i_1, i_2\}$ of f, there exists a morphism $F_1 \to F_2$, and we say that the map f is versal if it, along with the identity maps of source and target, is a versal unfolding of itself.

Theorem 3. [GWdPL76, III Theorem 3.4] *A smooth germ f is stable if and only if it is versal.* □

1.1.4. *Equivalences of maps and germs.* We have already defined the group $\mathscr{A}$ of equivalences of maps, and we can give a similar definition for germs; the group of

$\mathscr{A}$-equivalences of germs $(N, S) \to (P, y)$ is the product of groups of diffeomorphism germs of source and target:

$$\mathscr{A} = \mathrm{Diff}^{\infty}(N, S) \times \mathrm{Diff}^{\infty}(P, y),$$

where $\mathrm{Diff}^{\infty}(N, S)$ denote diffeomorphism germs which leave the points of the set S fixed.

When working with topological stability, $\mathscr{A}$-equivalence can sometimes be too much to ask for, and we introduce the concept of $\mathscr{A}_0$-equivalence. We say that two maps $f, g\colon N \to P$ are $\mathscr{A}_0$-equivalent if there exist homeomorphisms Ψ and Φ such that the diagram

$$\begin{array}{ccc} N & \xrightarrow{f} & P \\ {\scriptstyle\Psi}\downarrow & & \downarrow{\scriptstyle\Phi} \\ N & \xrightarrow[g]{} & P \end{array}$$

commutes. An analogous definition can be made for germs.

We are going to use other equivalences as well. We define $\mathscr{R}$ and $\mathscr{L}$ to be the groups $\mathrm{Diff}^{\infty}(N, S)$ and $\mathrm{Diff}^{\infty}(P, y)$ of diffeomorphism germs of source and target, respectively, giving $\mathscr{A} = \mathscr{R} \times \mathscr{L}$. We also define the group of $\mathscr{K}$-equivalences on the space of germs $(N, S) \to (P, y)$ as the group of diffeomorphism germs

$$H\colon (N \times P, S \times y) \to (N \times P, S \times y)$$

such that the diagram below commutes:

$$\begin{array}{ccccc} (N, S) & \xrightarrow{\mathrm{id} \times y} & (N \times P, S \times y) & \xrightarrow{\mathrm{pr}_N} & (N, S) \\ {\scriptstyle H_y}\downarrow & & {\scriptstyle H}\downarrow & & \downarrow{\scriptstyle H_y} \\ (N, S) & \xrightarrow[\mathrm{id} \times y]{} & (N \times P, S \times y) & \xrightarrow[\mathrm{pr}_N]{} & (N, S) \end{array}$$

where $H_y = H | N \times \{y\}$. Again, we ask that points of the set S stays fixed. Note, in particular, that $\mathscr{A} < \mathscr{K}$.

An element $H \in \mathscr{K}$ acts on a germ $f\colon (N, S) \to (P, y)$ by $H \cdot f = g$, where

$$(\mathrm{id}, f) \circ H_y = H \circ (\mathrm{id}, g)$$

as maps $(N, S) \to (N \times P, S \times y)$. There is a subgroup $\mathscr{C}$ of $\mathscr{K}$ consisting of elements H such that $H_y = \mathrm{id}$. In fact, $\mathscr{K} = \mathscr{C} \rtimes \mathscr{R}$, and for an element $(H, \psi) \in \mathscr{C} \rtimes \mathscr{R}$, you can think of its $\mathscr{K}$-action on f as

$$((H, \psi) \cdot f)(x) = \mathrm{pr}_P(H(x, f(\psi^{-1}(x)))).$$

Theorem 4. [GWdPL76, Chapter III Theorem 4.3] *Two germs*

$$f_0, f_1\colon (N, S) \to (P, y)$$

that admit stable unfoldings, are $\mathscr{K}$-equivalent if and only if their stable unfoldings of equal dimension are $\mathscr{A}$-equivalent. In particular, two stable maps are $\mathscr{K}$-equivalent if and only if they are $\mathscr{A}$-equivalent. □

1.1.5. *Jet space.* We define an equivalence relation on the set of germs $(\mathbb{R}^n, 0) \to (\mathbb{R}^p, 0)$ by saying that two germs f, g are equivalent if the degree $\leq k$ terms of their Taylor expansions coincide, for a fixed number $k \in \mathbb{N}$. The equivalence class of a germ f is denoted $j^k f$, and we call it the *k-jet of f*. We denote by $J^k(n,p)$ the set of k-jets of germs $(\mathbb{R}^n, 0) \to (\mathbb{R}^p, 0)$, which admits an obvious manifold structure (diffeomorphic to euclidean space) by projecting onto the coefficients of the Taylor polynomial. Given a source N and target P, we denote by $J^k(N,P)$ the set of k-jets of germs of maps $N \to P$; it is a manifold with various fiber bundle structures, for instance with bundle projection defined by projecting onto source and target, or forgetting higher order derivatives. For more information, see the book by Guillemin and Golubitsky [GG73].

Given a smooth map $f\colon N \to P$, we define a map $j^k f\colon N \to J^k(N,P)$ by setting $j^k f(x) = j^k(\hat{f}_x)$.

1.1.6. *Finite determinacy.* Let $\mathscr{H}$ be an equivalence of germs, such as $\mathscr{A}$ or $\mathscr{K}$. We say that a germ $f\colon (M,S) \to (N,y)$ is *$\mathscr{H}$-k-determined* if any other germ g with $j^k g = j^k f$ is $\mathscr{H}$-equivalent to f. Similarly, we say that a jet $z \in J^k(n,p)$ is *$\mathscr{H}$-sufficient* if any two germs g, f with k-jet z are $\mathscr{H}$-equivalent.

We say that germs which are $k - \mathscr{H}$-determined for some $k \in \mathbb{N}$, are *finitely $\mathscr{H}$-determined.* It is trivial that maps which are finitely $\mathscr{A}$-determined, are also finitely $\mathscr{K}$-determined.

For a germ $f\colon (N,S) \to (P,y)$, we define $\mathscr{N}_f = \theta_f/(tf(\theta_S) + f^* m_y \theta_f)$ and $d_e(f,\mathscr{K}) = \dim_{\mathbb{R}} \mathscr{N}_f$. Similarly, we define $d_e(f,\mathscr{A}) = \dim(\theta_f/tf(\theta_S) + wf(\theta_y))$.

Theorem 5 (Determinacy theorem). *For any map-germ*

$$f\colon (N,S) \to (P,y)$$

and group $\mathscr{H} = \mathscr{K}$ or $\mathscr{A}$, the following are equivalent:

i) *f is finitely $\mathscr{H}$-determined, and*
ii) $d_e(f,\mathscr{H}) < \infty$.
iii) *for some $k \in \mathbb{N}$, $m_y^k \theta_f \subset T\mathscr{H}(f)$, where*

$$T\mathscr{H}(f) = \begin{cases} tf(\theta_S) + f^* m_y \theta_f & \text{if } \mathscr{H} = \mathscr{K}, \\ tf(\theta_S) + wf(\theta_y) & \text{if } \mathscr{H} = \mathscr{A}. \end{cases}$$

If $\mathscr{H} = \mathscr{K}$ and iii) *holds for some k, then f is $k - \mathscr{K}$-determined.*

Proof. See [Mat68a, Theorem 3.6], and [Wal81, Theorem 1.2]. □

Proposition 6. *If $d_e(f,\mathscr{K}) = r - 1$, then f is $r - \mathscr{K}$-determined.*

Proof. If $d_e(f,\mathscr{K}) < r$, that is $\dim_{\mathbb{R}}(\theta_f/tf(\theta_S) + f^* m_y \theta_f) < r$, then certainly

$$\dim_{\mathbb{R}}(\theta_f/tf(\theta_S) + f^* m_y \theta_f + m_y^{r+1}\theta_f) < r,$$

and by [GWdPL76, Chapter IV, Lemma 1.2],

$$m_y^r\theta_f \subset tf(\theta_S) + f^*m_y\theta_f,$$

and by Theorem 5, f is $r - \mathscr{K}$-determined. □

Comparing with the definition of infinitesimal stability it is easy to see that stable germs are finitely $\mathscr{A}$-determined. Furthermore, we have:

Theorem 7. *Let $f\colon (N,S) \to (P,y)$ be a smooth germ. The following conditions are equivalent:*

i) *f is finitely $\mathscr{K}$-determined,*
ii) *f has a stable unfolding.*

Proof. This follows from [GWdPL76, III, Theorem 2.8] and [dPW95, Theorem 2.2.1]. □

Definition 8. [Finite singularity type] We say that a map $f\colon N \to P$ has *finite singularity type* (FST) if the following conditions hold:

i) $f|\Sigma f$ is proper,
ii) the cardinalities of the sets $\Sigma F \cap f^{-1}(y)$ for $y \in P$ are uniformly bounded, and
iii) the codimensions $d_e(\hat{f}_x, \mathscr{K})$ for $x \in N$ are uniformly bounded.

Given a jet $z \in J^k(n,p)$, choose a representative $\hat{f}$, define

$$\mathscr{N}_z = \mathscr{N}_{\hat{f}}/m_x^{k+1}\theta_{\hat{f}}$$

and

$$d_e(z,\mathscr{K}) = \dim_{\mathbb{R}} \mathscr{N}_z.$$

Using Nakayama's lemma one can show that $d_e(z,\mathscr{K}) < k$ implies $d_e(\hat{f},\mathscr{K}) < k$, ensuring that $\hat{f}$ is $k - \mathscr{K}$-determined by Proposition 6, and that z is $\mathscr{K}$-sufficient.

Theorem 9. *The set of maps in $C^\infty_{\text{proper}}(N,P)$ which have FST coincides with the set of maps that admit stable unfoldings, and this set is dense in $C^\infty_{\text{proper}}(N,P)$.*

Proof. Define $W^k(n,p) = \{z \in J^k(n,p) | d_e(z,\mathscr{K}) \geq k\}$. For k sufficiently large, the set of maps in $C^\infty_{\text{proper}}(N,P)$ whose jet extension $N \to J^k(N,P)$ avoid the corresponding subset $W^k(N,P) \subset J^k(N,P)$ is dense [Tou71]. All the (multi)germs of such maps are $k - \mathscr{K}$-determined. By [dPW95, Theorems 3.5.4 and 3.5.6], these maps are exactly those that admit stable unfoldings. □

1.1.7. *Local algebras and $\mathscr{K}$-equivalence.* One way of characterizing $\mathscr{K}$-equivalent germs $f\colon (N,S) \to (P,y)$ is through their *local algebras*.

The local algebra of f is defined as $Q(f) = \mathscr{E}_S/f^*m_y \cdot \mathscr{E}_S$, where $\mathscr{E}_S$ denotes the set of function germs $(M,S) \to \mathbb{R}$, m_y denotes the set of function germs

$(N, y) \to (\mathbb{R}, 0)$ and f^* denotes the map $\mathscr{E}_y \to \mathscr{E}_S$, $\phi \mapsto \phi \circ f$. If we decompose $f = \bigsqcup_{i=1}^{s} f_i \colon \bigsqcup_s (\mathbb{R}^n, 0) \to (\mathbb{R}^p, 0)$ then

$$Q(f) = \bigoplus_s \mathscr{E}(n) / \left(\bigoplus_{i=1}^{s} f_i^* m(p) \cdot \mathscr{E}(n) \right) = \bigoplus_{i=1}^{s} Q(f_i)$$

Theorem 10. [Mat69a, Theorem 2.1]

i) *If F is an unfolding of f, then $Q(F) \cong Q(f)$.*
ii) *Two finitely $\mathscr{K}$-determined map-germs $(N, S) \to (P, y)$ are $\mathscr{K}$-equivalent if and only if their local algebras are isomorphic.* □

We say that two germs f and g are $\mathscr{E}\mathscr{K}$-equivalent if their local algebras are isomorphic. Then two $\mathscr{E}\mathscr{K}$-equivalent germs admitting stable unfoldings have $\mathscr{A}$-equivalent unfoldings. In particular, two germs are $\mathscr{E}\mathscr{K}$-equivalent if they have $\mathscr{K}$-equivalent suspensions (where $f \times \mathrm{id}_{\mathbb{R}^d}$ is the d-dimensional suspension of f).

Using the above, we obtain a standard way to choose coordinates for stable multigerms:

Lemma 11. *Let $f \colon (N, \{x_1, \ldots, x_s\}) \to (P, y)$ be a stable multigerm. Then there is a change of coordinates (Ψ, Φ) as shown in the diagram below:*

$$\begin{array}{ccc} (N, \{x_1, \ldots, x_s\}) & \xrightarrow{\quad f \quad} & (P, y) \\ \Big\downarrow \Psi & & \Big\downarrow \Phi \\ \bigsqcup_{i=1}^{s} \left(s(F_i) \times \prod_{j \neq i} t(F_j) \times \mathbb{R}^d, 0 \right) & \xrightarrow[F]{} & \left(\prod_{j=1}^{s} t(F_j) \times \mathbb{R}^d, 0 \right) \end{array}$$

where $F = \bigsqcup_{i=1}^{s} \sigma_i \circ (F_i \times \mathrm{id}_{\prod_{j \neq i} t(F_j) \times \mathbb{R}^d})$, F_i is a ministable germ $\mathscr{E}\mathscr{K}$-equivalent to the germ of f at x_i, σ_i moves the 1st coordinate to the ith in $\prod_{i=1}^{s} t(F_i)$, and $d \in \mathbb{N}_0$ is chosen to get the appropriate dimensions.

Proof. Since the F_i are $\mathscr{E}\mathscr{K}$-equivalent to the f_i, the local algebras $Q(F_i)$ and $Q(f_i)$ are isomorphic; and as a consequence also the local algebras $Q(F)$ and $Q(f)$ are isomorphic, all of this by Theorem 10. Hence F and f are $\mathscr{K}$-equivalent. But F and f are also both stable, hence by Theorem 4, F and f are $\mathscr{A}$-equivalent. □

1.1.8. *Topological stability.* A map $f \in C^\infty(N, P)$ is *topologically stable* if there exists a neighborhood $\mathscr{U}$ of f in the strong topology on $C^\infty(N, P)$ such that each $g \in \mathscr{U}$ is $\mathscr{A}_0$-equivalent to f via a pair of homeomorphisms $(\Psi, \Phi) \in \mathscr{A}_0$. If we can choose the (Ψ, Φ) to depend continuously on $g \in \mathscr{U}$, then we say that f is *strongly* topologically stable.

For a discussion of other "flavors" of topological stability, we refer to Chapter 2.1 and the book by du Plessis and Wall [dPW95].

1.1.9. *How it all fits together.* Let us summarize:

Theorem 12. *Let $f\colon N \to P$ be a smooth, quasi-proper map (i.e. there exists a neighborhood U of Σf in N such that $f|U$ is proper). The following conditions are equivalent:*

i) *f is stable,*
ii) *f is infinitesimally stable,*
iii) *f is locally stable,*
iv) *f is versal,*
v) *given a parametrized unfolding F of f, we can find a smooth retraction $F \to f$.*

Proof. These are well-known facts, mostly originating in Mather's classic papers [Mat69b, Mat68a, Mat69a]. See also [dPW95, Chapters 2, 3, and 4]. □

We can prove similar relations for stability of germs.

Theorem 13. *Let $f\colon (N, S) \to (P, y)$ be a smooth germ; then the following conditions are equivalent:*

i) *f is infinitesimally stable,*
ii) *f is versal,*
iii) *given any unfolding F of f, we can find a smooth retraction $F \to f$.*

Proof. These are also due to Mather, see [Mat69b, Mat68a, Mat69a] and the book by du Plessis and Wall [dPW95, Chapters 2, 3 and 4]. □

1.2. **Differential topology.** We will use a great deal of differential topology, most of which we assume the reader to be familiar with. Some of the constructions which we will use are, however, not so commonly known, and we give their basic definitions and properties, along with references.

1.2.1. *Tubular neighborhoods and sprays.* The concept of a tubular neighborhood is well-known, and tubular neighborhoods are usually constructed using normal bundles [Hir76, Lee03]. However, there is a less known construction of tubular neighborhoods using *sprays*, which we shall dwell on simply because it deserves attention. For details, we refer to [Lan95, Chapter IV].

Suppose that M is a smooth submanifold of N, and that we wish to construct a tubular neighborhood of M in N. Rather than construct the tubular neighborhoods using vector fields $M \to TN$ normal to M, the spray method uses vector fields *on the tangent bundle of* N, and pulls the resulting smooth foliation by fibers back to N using an exponential map. The advantage of this technique is that the conditions for a spray are convex, and as a consequence, sprays can be glued together using partitions of unity.

Given a smooth manifold N, we denote by $\pi\colon TN \to N$ the tangent bundle projection. A *second order vector field* on N is a vector field X on TN such that

$T\pi \circ X = \mathrm{id}.$

$$\begin{array}{ccc} TN & \xrightarrow{\;X\;} & TTN \\ & \overset{\mathrm{id}}{\searrow} & \downarrow {\scriptstyle T\pi} \\ & & TN \end{array}$$

Let $p\colon E \to B$ be any vector bundle, and let $s \in \mathbb{R}$. There is a vector bundle morphism $s_E\colon E \to E$ given by $(x, v) \mapsto (x, sv)$ for all $x \in B$ and $v \in E_x$. If $s \neq 0$, this map is a vector bundle isomorphism.

A *spray* on N is a second order vector field X on N such that

$$X(sv) = Ts_{TN}(sX(v)), \qquad \forall\, s \in \mathbb{R},\ v \in TN.$$

Remark 14. Both the second order vector field condition and the spray condition are convex; hence we may combine sprays using partitions of unity.

Given a spray X on N, let $v \in TN$ and let β_v be the integral curve of X with $\beta_v(0) = v$. Let $\mathscr{D}$ be the set of vectors $v \in TN$ such that β_v is defined on $[0, 1]$; now $\mathscr{D}$ is a neighborhood of the zero section in TN, and the map

$$v \mapsto \beta_v(1)$$

is a morphism $\mathscr{D} \to TN$. Define the exponential map as the smooth map given by

$$\exp\colon \mathscr{D} \to N,\ \exp(v) = \pi(\beta_v(1)). \tag{15}$$

Note in particular that for each $x \in N$, we get $\exp(0_x) = x$, where $0_x \in T_xN$.

Return now to the problem of constructing a tubular neighborhood of $M \subset N$. Consider the short exact sequence

$$0 \to TM \to TN|M \to NM \to 0.$$

This sequence splits, and $TN|M = TM \oplus NM$. Given a spray X on N, we construct the corresponding exponential map, and denote its restriction to NM by $\exp|N\colon \mathscr{D} \cap NM \to N$.

Proposition 16. [Lan95, Chapter IV, Theorem 5.1] *The map* (15) *is a local isomorphism, and we can shrink $\mathscr{D} \subset NM$ to a neighborhood W of the zero section in NM, such that*

$$\exp|W\colon W \to N$$

is an isomorphism. In particular, it is a tubular neighborhood embedding. □

We have thus seen how to construct a tubular neighborhood from a given spray. Suppose, on the other hand, that we are given a tubular neighborhood (T, π) of $M \subset N$, and that $x \in M$. We show how to construct a spray X in a neighborhood U of x in N, such that X induces the tubular neighborhood $(T, \pi)|U$.

Using a Euclidean chart at x we may assume that U is an open neighborhood of $x = 0$ in $\mathbb{R}^n$, and that $M \cap U = \mathbb{R}^m \times \{0\} \subset \mathbb{R}^m \times \mathbb{R}^{n-m} = \mathbb{R}^n$.

We can "straighten out" the fibers of π (see Lemma 24) such that the bundle projection π is just the Euclidean projection $\mathbb{R}^n = \mathbb{R}^m \times \mathbb{R}^{n-m} \to \mathbb{R}^m$, restricted to U.

The tangent bundle of $\mathbb{R}^n$ is just $\mathbb{R}^n \times \mathbb{R}^n$, and $TT(\mathbb{R}^n) = (\mathbb{R}^n \times \mathbb{R}^n) \times (\mathbb{R}^n \times \mathbb{R}^n)$. The vector field $T\mathbb{R}^n \to TT\mathbb{R}^n$ given by $(v, w) \mapsto (v, w, w, 0)$ is a spray, and it defines the tubular neighborhood of $\mathbb{R}^n$ whose fibers are just the $\{\tilde{x}\} \times \mathbb{R}^{n-m}$, i.e. whose projection is $\mathbb{R}^m \times \mathbb{R}^{n-m} \to \mathbb{R}^m$. But this defines our original tubular neighborhood $(T, \pi)|U$.

Using this, and the fact that sprays can be combined using partitions of unity, we can start out with a collection of local tubular neighborhoods, whose properties near special points we want to preserve – and create a new, global tubular neighborhood with the desired properties near the special points.

1.2.2. *Stratifications.* We say that $F\colon N \to P$ is a stratified map if there are stratifications $\mathscr{S}'$ and $\mathscr{S}$ on N and P such that for any stratum $S' \in \mathscr{S}'$ there exists a stratum $S \in \mathscr{S}$ such that $F(S') \subset S$. If $F|S'$ is smooth for each stratum $S' \in \mathscr{S}'$, then F is stratified smooth.

For a stratification $(\mathscr{S}', \mathscr{S})$ of a map F of FST to be *ST-invariant* (ST being short for *stably-topologically*) means that if $y_1 \in S$, $y_2 \notin S$ for some $S \in \mathscr{S}$, then no two stable unfoldings of the germs $\hat{F}_{y_1}$, $\hat{F}_{y_2}$ can ever be $\mathscr{A}_0$-equivalent. The stratifications appearing in this thesis will all be ST-invariant.

Suppose that M is a smooth manifold with a stratification $\mathscr{S}$ by smooth submanifolds. We say that a vector field $\xi\colon M \to TM$ is *stratified smooth* if its restriction $\xi|\colon S \to TM$ to any stratum $S \in \mathscr{S}$ is smooth and tangent to the stratum.

In this thesis, stratifications will appear as partitions by presented singularity types. We denote by $\prod \Delta_i$ the germ class represented by a multigerm

$$\bigsqcup_{i=1}^{s} f_i\colon \bigsqcup_{i=1}^{s} (\mathbb{R}^n, 0) \to (\mathbb{R}^p, 0),$$

where f_i belongs to the monogerm class Δ_i.

Given a particular $\mathscr{E}\mathscr{K}$-germ class Δ, we write $\Delta(f)$ for the presentation of Δ in the target of a map $f\colon N \to P$, namely the set of points $y \in P$ such that the germ $\hat{f}\colon (N, S) \to (P, y)$ belongs to Δ for some subset $S \subset f^{-1}(y) \cap \Sigma f$. We define the *strict* presentation $\Delta_{\text{strict}}(f)$ to be the subset of $\Delta(f)$ where the germ $\hat{f}_y\colon (N, f^{-1}(y) \cap \Sigma f) \to (P, y)$ belongs to Δ. We denote by $\Delta_{\text{source}}(f)$ the corresponding subset $\{f^{-1}(y) \cap \Sigma f | y \in \Delta_{\text{strict}}(f)\}$ of source.

2. E-tame retractions

Suppose that $F\colon (\mathbb{R}^n \times \mathbb{R}^d, 0) \to (\mathbb{R}^p \times \mathbb{R}^d, 0)$ is a smoothly stable, $\mathbb{R}^d$-level preserving germ which unfolds $f\colon (\mathbb{R}^n, 0) \to (\mathbb{R}^p, 0)$; then by Theorem 13 f is smoothly stable if and only if we can find a pair of smooth retractions $r\colon (\mathbb{R}^{n+d}, 0) \to (\mathbb{R}^n, 0)$ and $s\colon (\mathbb{R}^{p+d}, 0) \to (\mathbb{R}^p, 0)$ such that $f \circ r = s \circ F$.

Moreover, writing z for the projection $(\mathbb{R}^{p+d},0) \to (\mathbb{R}^d,0)$, z is submersive, as is the composition $z \circ F$, which is really just a projection onto $(\mathbb{R}^d,0)$ from the source. The map-germs $(r, z\circ F)$ and (s,z) are diffeomorphism germs which smoothly trivialize F over f; that is, $(f \times \mathrm{id}_{(\mathbb{R}^d,0)}) \circ (r, z \circ F) = (s,z) \circ F$.

The situation with respect to topological equivalence is not so straightforward, as being topologically equivalent to a topologically stable germ may not be enough to secure topological stability – we shall see in Section 2.1 that the proof from the smooth stability case does not carry over. du Plessis and Wall [dPW95] introduced the concept of *tame retractions*, where tameness refers to geometric conditions which are sufficient for retractions from smoothly stable unfoldings to preserve topological stability.

2.1. **Tame retractions and topological stability.** Tame retractions were first defined by du Plessis and Wall [dPW95, Chapter 4], and we follow their introduction, which builds on the following result on C^r-smooth stability:

Proposition 17. *Suppose that $f\colon N \to P$ is a smooth map of FST, and that N is compact. Suppose furthermore that any smooth d-parameter family of mappings ($1 \le d \le \infty$) containing f can be trivialized by level-preserving C^r-diffeomorphisms near f. More precisely, if $F \in C^\infty_{\mathrm{lp}}(N \times \mathbb{R}^d, P \times \mathbb{R}^d)$ such that $F_0 = f$, then there exist a neighborhood U of the origin in $\mathbb{R}^d$ and diffeomorphisms $R \in \mathrm{Diff}^r_{\mathrm{lp}}(N \times U)$ and $L \in \mathrm{Diff}^r_{\mathrm{lp}}(P \times U)$ such that*

$$L \circ (F|N \times U) = (f \times \mathrm{id}_U) \circ R.$$

Then f is strongly C^r-stable.

This is [dPW95, Proposition 4.2.2], and similar results can be proven for germs and for maps with non-compact source; the details can be found in the book [dPW95]. We sketch the proof of Proposition 17, as this will lead us to the definition of tame retractions.

Proof. Since f is of FST, it admits a proper, C^∞-stable unfolding F by Theorem 9 (see [dPW95, Theorem 3.5.6] for proper version), and since N is compact we may assume that F is level-preserving; more precisely $F \in C^\infty_{\mathrm{lp}}(N \times \mathbb{R}^d, P \times \mathbb{R}^d)$. By [dPW95, Theorem 2.4.5] there exist a neighborhood $\mathscr{U}$ of f in $C^\infty(N,P)$ and a continuous map $(i,j)\colon \mathscr{U} \to C^\infty(N, N \times \mathbb{R}^d) \times C^\infty(P, P \times \mathbb{R}^d)$ such that $\{F, i(g), j(g)\}$ unfolds g for all $g \in \mathscr{U}$, and $i(f), j(f)$ are the natural inclusions. By the assumptions on f, there exist a neighborhood U of 0 in $\mathbb{R}^d$ and diffeomorphisms $R \in \mathrm{Diff}^r_{\mathrm{lp}}(N \times U)$, $L \in \mathrm{Diff}^r_{\mathrm{lp}}(P \times U)$ such that

$$L \circ (F|N \times U) = (f \times \mathrm{id}_U) \circ R.$$

Now there is a neighborhood $\mathscr{V}$ of $\mathrm{id}_P \times 0_U$ in $C^\infty(P, P \times \mathbb{R}^d)$ such that for $\tilde{j} \in \mathscr{V}$, the composition $\mathrm{pr}_P \circ L \circ \tilde{j}$ is a C^r-diffeomorphism, since the map $C^\infty(P, P \times \mathbb{R}^d) \to C^r(P,P)$ defined by $\tilde{j} \mapsto \mathrm{pr}_P \circ L \circ \tilde{j}$ is continuous and the set $\mathrm{Diff}^r(P,P)$ is open in

$C^r(P, P)$. Since j is continuous, we may assume that $\mathrm{pr}_N \circ R \circ i(g)$ and $\mathrm{pr}_P \circ L \circ j(g)$ are C^r-diffeomorphisms when $g \in \mathscr{U}$. But then

$$(\mathrm{pr}_P \circ L \circ j(g)) \circ g = f \circ (\mathrm{pr}_N \circ R \circ i(g)),$$

and since $g \mapsto \mathrm{pr}_N \circ R \circ i(g)$ and $g \mapsto \mathrm{pr}_P \circ L \circ j(g)$ are continuous maps $\mathscr{U} \to \mathrm{Diff}^r(N)$ and $\mathscr{U} \to \mathrm{Diff}^r(P)$, the proposition holds. □

Note that if $r = 0$, on the other hand, we are not necessarily able to find a neighborhood of $\mathrm{id}_P \times 0_U$ such that for j in the neighborhood, $\mathrm{pr}_P \circ L \circ j$ is a homeomorphism, and consequently, we cannot necessarily find the neighborhood $\mathscr{U}$ of f such that $\mathrm{pr}_P \circ L \circ j(g)$ is a homeomorphism for all $g \in \mathscr{U}$.

The C^r diffeomorphism L corresponds to a retraction $\mathrm{pr}_P \circ L\colon P \times \mathbb{R}^d \to P$, and we say that a retraction s to an embedding $e\colon P \to P \times \mathbb{R}^d$ is *tame* whenever there exists a neighborhood of e in $C^\infty(P, P \times \mathbb{R}^d)$ such that all embeddings j in that neighborhood combine with s to form a homeomorphism $s \circ j$.

We also define a concept of retraction between unfoldings: Given an unfolding $F\colon N' \to P'$ of $f\colon N \to P$, we say that two retractions $r\colon N' \to N$ and $s\colon P' \to P$ define a retraction $(r, s)\colon F \to f$ if $s \circ F = f \circ r$ and the map $(F, r)\colon N' \to P' \times N$ is injective.

Following the proof of Proposition 17, we then see:

Proposition 18. *Suppose that $f\colon N \to P$ has FST, and that N is compact. Assume furthermore that for any unfolding $\{F, i, j\}$ of f there exist a neighborhood V of $j(P)$ and a tame retraction $(r, s)\colon F|F^{-1}(V) \to f$; that is, a retraction (r, s) with s tame. Then f is strongly C^0-stable.* □

Remark 19.
i) In particular, any C^1 retraction $r\colon M' \to M$ is tame. As is mentioned in the proof above, the diffeomorphisms form a neighborhood of id_M in $C^1(M, M)$, and the map $r_*\colon C^k(M', M) \to C^1(M, M)$ given by $r_*(\phi) = r \circ \phi$ is continuous.
ii) Tame retractions induce topological triviality. More precisely, if there is an E-tame retraction $F \to f$, then F is $\mathscr{A}_0$-equivalent to $f \times \mathrm{id}_{\mathbb{R}^d}$, as we soon shall see in Lemma 24.

We use a stronger idea of tameness, namely the "extremely tame" or, for short, *E-tame* retractions defined by du Plessis and Wall [dPW95]. These retractions satisfy some additional geometric conditions and have nice functorial properties, in the sense that E-tameness is often preserved when we induce new retractions from a given one.

Definition 20. [$C^{0,1}$-foliations and E-tame retractions] A $C^{0,1}$-foliation of a smooth n-manifold N is a partition $\mathscr{F}$ of N such that for any $y \in N$ there exist open neighborhoods W of y in N and U, V of 0 in $\mathbb{R}^m$ and $\mathbb{R}^{n-m}$, respectively, and a homeomorphism $\phi\colon U \times V \to W$ such that for each $u \in U$ there exists $F \in \mathscr{F}$ such that $\phi(u \times V) = W \cap F$, each leaf F is a C^1 submanifold of N, and the tangent space T_yF varies continuously with $y \in N$.

Suppose $i\colon M \to M'$ is an embedding. We say that the retraction $r\colon M' \to M$ to i is *E-tame* if its fibers form a $C^{0,1}$-foliation transverse to $i(M)$, in a neighborhood of $i(M)$. A germ of retraction is E-tame if it has an E-tame representative.

Let $f\colon M \to N$; let $\{F\colon M' \to N'; i\colon M \to M', j\colon N \to N'\}$ be an unfolding of f, and let $(r,s)\colon F \to f$ be a retraction to (i,j), that is, $s \circ F = f \circ r$ and $(r,F)\colon M' \to M \times N'$ is injective. We say that (r,s) is *E-tame* if s is E-tame.

Proposition 21. *An E-tame retraction is tame.*

Proof. This follows from [dPW95, Proposition 9.3.3]. □

In Proposition 18 we saw that a map is strongly topologically stable if we can always retract tamely onto it from a stable unfolding. In fact, it is enough to find *one* E-tame retraction from a stable germ in order to prove topological stability, that is, E-tame retractions from stable maps preserve topological stability:

Theorem 22. *Suppose that $(r,s)\colon F \to f$ is an E-tame retraction, where*

$$\{F\colon (\mathbb{R}^{n+a},0) \to (\mathbb{R}^{p+a},0), i, j\}$$

is a stable unfolding of $f\colon (\mathbb{R}^n,0) \to (\mathbb{R}^p,0)$.

Then, for any other stable unfolding $\{\tilde{F}\colon \mathbb{R}^{n+b} \to \mathbb{R}^{p+b}; \tilde{i}, \tilde{j}\}$ of f, we can find an E-tame retraction $(\tilde{r},\tilde{s})\colon \tilde{F} \to f$; so f satisfies the conditions of Proposition 18, and in particular, f is strongly topologically stable.

Remark 23. Note that the theorem also holds for maps, with similar assumptions as for Proposition 18. The theorem also holds for the other versions of tameness defined by du Plessis and Wall [dPW95].

Proof. We write the proof for germs; the same proof can be adapted even in the various map cases.

Given suitable $c, d \in \mathbb{N}_0$ (at least one of which can be assumed to be 0), $\tilde{F} \times \mathrm{id}_{(\mathbb{R}^c,0)}$ and $F \times \mathrm{id}_{(\mathbb{R}^d,0)}$ are smoothly equivalent by diffeomorphism germs Φ and Ψ, as the two germs are stable unfoldings of f. Hence we obtain an E-tame retraction $\tilde{F} \times \mathrm{id}_{(\mathbb{R}^c,0)} \to f$. But now, if we denote by i_0 and j_0 the zero level embeddings of $\mathbb{R}^{n+b}$ and $\mathbb{R}^{p+b}$ into $\mathbb{R}^{n+b+c}$ and $\mathbb{R}^{p+b+c}$, we see that $(r \circ i_0, s \circ i_0)$ are E-tame retractions $\tilde{F} \to f$. □

2.2. Properties of E-tame retractions.

As in the smooth case, there are relations between E-tame retractions and triviality.

Lemma 24. *Suppose that $(r,s)\colon F \to f$ is an E-tame retraction (between maps or germs), where $f\colon N \to P$ and $F\colon N \times U \to P \times U$ is a U-level preserving unfolding of f. Then there exist neighborhoods W and V of $N \times \{0\}$ and $P \times \{0\}$ in $N \times U$ and $P \times U$ such that $F(W) \subset V$ and $F|W$ is topologically equivalent to $(f \times \mathrm{id}_U)|W$; more specifically, $(r, \mathrm{pr}_U)|W$ and $(s, \mathrm{pr}_U)|V$ are open embeddings, and*

$$(s, \mathrm{pr}_U) \circ F|W = (f \times \mathrm{id}) \circ (r, \mathrm{pr}_U)|W.$$

Furthermore, (s, pr_U) *restricted to any fiber of* s *is a* C^1*-diffeomorphism onto its image. Finally, if* s *[and* r*] is smooth, then* (s, pr_U) *[and* (r, pr_U)*] is a diffeomorphism.*

Proof. We identify N and P with the zero-level inclusions $N \times \{0\}$ and $P \times \{0\}$ in $N \times U$ and $P \times U$. We start by proving the claims for (s, pr_U). First we show that if V is sufficiently small, then $\mathrm{pr}_U \colon V \cap s^{-1}(y, 0) \to U$ is injective for any $(y, 0) \in P \times U$.

If this is not the case, then we can find a point $(y, 0) \in P \times \{0\}$ and sequences $(y_i, u_i), (\tilde{y}_i, u_i) \in V$ such that $s(y_i, u_i) = s(\tilde{y}_i, u_i)$ for all i, $y_i \neq \tilde{y}_i$, and $(y_i, u_i), (\tilde{y}_i, u_i) \to (y, 0)$. Now $v := \lim[(y_i, u_i) - (\tilde{y}_i, u_i)] = \lim[(y_i - \tilde{y}_i, 0)] \in T_{(y,0)}P$ (where $[v]$ denotes the unit vector parallel to a vector v), but this is impossible because $v \in T_{(y,0)}s^{-1}(y, 0)$, and $s^{-1}(y, 0)$ is transverse to P and of complementary dimension. Thus we may assume that $(s, \mathrm{pr}_U)|V$ is injective, and by invariance of domain it will be an open embedding.

Next we show that the projection $\mathrm{pr}_U \colon V \to U$ restricts to a submersion on the fibers of s when V is sufficiently small. If this is not the case, then there exist $(y, 0) \in P \times \{0\}$ and a sequence $(y_i, u_i) \in P \times U$ converging towards $(y, 0)$, and there exists $0 \neq v_i \in T_{(y_i,u_i)}\mathscr{F}$ such that $\|v_i\| = 1$ and $\mathrm{pr}_U v_i = 0$. Here $\mathscr{F}$ denotes the foliation by fibers of s, and $T_z\mathscr{F}$ denotes the tangent space at z to the leaf of $\mathscr{F}$ containing z. By passing to a subsequence, we may assume $v_i \to v$ for some unit vector v since the v_i all belong to the compact unit circle. But then there exists, for each $i \in \mathbb{N}$, a sequence $(y_i^j, u_i^j)_{j \in \mathbb{N}}$ such that $s(y_i^j, u_i^j) = s(y_i, u_i)$ and

$$\lim_{j \to \infty} [(y_i^j, u_i^j) - (y_i, u_i)] = v_i.$$

Now we can define a sequence $(\tilde{y}_i, \tilde{u}_i)$ such that $(\tilde{y}_i, \tilde{u}_i)$ is one of the (y_i^j, u_i^j) for each i, $(\tilde{y}_i, \tilde{u}_i) \to (y, 0)$ and $[(\tilde{y}_i, \tilde{u}_i) - (y_i, u_i)] \to v$. In particular, $s(\tilde{y}_i, \tilde{u}_i) = s(y_i, u_i)$ for each i so $v \in T_{(y,0)}\mathscr{F}$. But this is impossible, since by continuity, $\mathrm{pr}_U(v) = \mathrm{pr}_U(\lim v_i) = \lim \mathrm{pr}_U(v_i) = 0$. Hence we may assume that $\mathrm{pr}_U|V \cap s^{-1}(y, 0)$ is submersive. Being submersive smooth homeomorphisms, the restrictions of Φ to the fibers of s are diffeomorphisms.

Next, we turn to r, and prove that (r, pr_U) is an open embedding – the rest of the claims will follow by the same argument as for s. By the invariance of domain, it is enough to show that (r, pr_U) is injective. Recall that by the definition of a retraction $F \to f$, the map $(F, r) \colon N \times U \to P \times U \times N$ is injective.

Suppose that $(r, \mathrm{pr}_U)(x, u) = (r, \mathrm{pr}_U)(\tilde{x}, \tilde{u})$ for $(x, u), (\tilde{x}, \tilde{u}) \in N \times U$. Then

$$\begin{aligned}
& u = \tilde{u} \text{ and } r(x, u) = r(\tilde{x}, \tilde{u}) \\
\Rightarrow\ & u = \tilde{u} \text{ and } s(F(x, u)) = f(r(x, u)) = f(r(\tilde{x}, \tilde{u})) = s(F(\tilde{x}, \tilde{u})) \\
\Rightarrow\ & (s, \mathrm{pr}_U)(F(x, u)) = (s, \mathrm{pr}_U)(F(\tilde{x}, \tilde{u})) \\
\Rightarrow\ & F(x, u) = F(\tilde{x}, \tilde{u}) \\
\Rightarrow\ & (F, r)(x, u) = (F, r)(\tilde{x}, \tilde{u}) \\
\Rightarrow\ & (x, u) = (\tilde{x}, \tilde{u}).
\end{aligned}$$

This completes the proof. □

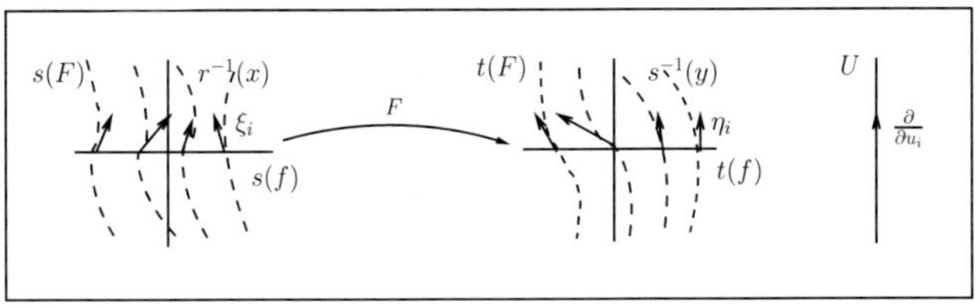

FIGURE 1. Given an E-tame retraction $(r, s)\colon F \to f$, we can find integrable vector fields η_i in target, tangent to the fibers of s, such that η_i lifts $\frac{\partial}{\partial u_i}$ over pr, and such that the flows of η_i induce s as described in (25). If r has C^1-smooth fibers and we can find integrable vector fields ξ_i in source, tangent to the fibers of r, lifting η_i over F, then the flows of the ξ_i will induce r as described in (25). This happens, for instance, if r is also E-tame (and, in particular, if r is smooth).

2.2.1. *E-tame retractions induced by vector fields.* Let (r, s) be a germ of E-tame retraction

$$\{F\colon (N', x) \to (P', y)\} \to \{f\colon (N, x) \to (P, y)\}$$

as in Definition 20. It follows from Lemma 24 that we may pick a submersion $z\colon (P', y) \to (\mathbb{R}^d, 0)$, such that $z|s^{-1}(y)$ is a C^1 diffeomorphism-germ onto $(\mathbb{R}^d, 0)$ for each $y \in P$, and such that $z^{-1}(0) = (P, y')$. If the fibers of r are smooth and transverse to N, we see that even $z \circ F|r^{-1}(x)$ is a C^1 diffeomorphism-germ, and that $(z \circ F)^{-1}(0) = (N, x')$.

We can find vector fields $\{\eta_i\}$ on N' and $\{\xi_i\}$ on M' which are tangent to and C^1 smooth on the fibers of s and r (assuming that the fibers of r are smooth), satisfying the following conditions: If $\frac{\partial}{\partial u_i}$ denotes the standard i^{th} coordinate vector field on $\mathbb{R}^d$, then η_i is a lift of $\frac{\partial}{\partial u_i}$ over z, and ξ_i is a lift of η_i over F. If the ξ_i are integrable, we denote their flows by Ψ_i. The flow of η_i is denoted Φ_i (η_i is continuous, and hence integrable, since the foliation is $C^{0,1}$). Then we have

$$\begin{aligned} s(y) &= \Phi_d(\ldots(\Phi_1(y, -z_1(y)), \ldots, -z_d(y)) \\ r(x) &= \Psi_d(\ldots(\Psi_1(x, -(z \circ F)_1(x), \ldots, -(z \circ F)_d(x)). \end{aligned} \tag{25}$$

See Figure 1.

Since s is E-tame, the vector fields η_i will be continuous. This, however, will not be the case for the ξ_i, unless r is also E-tame. If r is not E-tame, there is also no guarantee that the vector fields have a continuous global flow, although they have (possibly not unique) integral curves.

The vector fields are not uniquely defined, and for a given retraction, they will depend on the choice of submersion z.

The converse procedure is even more complicated. Although families of integrable vector fields ξ_i and η_i *will* define retractions through the formula (25), they do not have to be E-tame, or even have smooth fibers, and even if they do, we are not guaranteed that the original vector fields are tangent to the fibers of the final retraction.

If the vector fields span $C^{0,1}$-foliations in source and target which are transverse to (M, N), then the vector fields generate E-tame retractions. This leads us to the problem of integrability of continuous distributions. In fact, the sets of vector fields $\{\xi_i\}_{i=1}^d$ and $\{\eta_i\}_{i=1}^d$ span a $C^{0,1}$-foliation if and only if the corresponding sets $\mathbb{R}\{\xi_i\}_{i=1}^d$ and $\mathbb{R}\{\eta_i\}_{i=1}^d$ form a Lie algebra, i.e. if they are closed under the Lie bracket. In particular, if $d = 1$, they induce an E-tame retraction.

There are examples of sets of continuous, integrable and even controlled vector fields for which the induced retraction is not tame [dPW95, Remark 9.3.14].

If, however, the vector fields are smooth, then the retractions defined by the formula (25) will be smooth, and thus they will be E-tame. But the defining vector fields need not be tangent to the fibers of the resulting retraction, i.e. the resulting foliation need not be spanned by the original vector fields.

2.2.2. *Properties.* We defined tameness to mean that a tame retraction combines into a homeomorphism with nearby embeddings, and this gives rise to the following property of E-tame retractions: Suppose that we have an E-tame retraction $(r, s)\colon F \to f$ and a germ g also unfolded by F, whose target (viewed as a subset of the target of F) is transverse to the fibers of s. Then we can construct another E-tame retraction $(r', s')\colon F \to g$ whose fibers in source and target of F coincide with the fibers of (r, s), by sliding along the fibers onto the respective submanifolds.

Proposition 26. *Suppose that we are given a multigerm* $f\colon \bigsqcup_k(\mathbb{R}^n, 0) \to (\mathbb{R}^p, 0)$, *an unfolding* $\{F\colon \bigsqcup_k(\mathbb{R}^{n+d}, 0) \to (\mathbb{R}^{p+d}, 0); i_1, j_1\}$ *of* f, *and an E-tame retraction* $(r, s)\colon F \to f$.

Suppose that $g\colon \bigsqcup_k(\mathbb{R}^n, 0) \to (\mathbb{R}^p, 0)$ *is another multigerm such that the triple* $\{F; i_2, j_2\}$ *unfolds* g, *and such that the foliation* $\mathscr{F}$ *corresponding to* s *is transverse to* $j_2(\mathbb{R}^p, 0)$. *Then*

- i) $r \circ i_2$ *and* $s \circ j_2$ *are homeomorphism germs, and*
- ii) *we can find an E-tame retraction* $(R, S)\colon F \to g$, *induced by* (r, s).

If the E-tame retraction (r, s) *is stratified smooth with respect to an ST-invariant stratification of* F, *then so is the new E-tame retraction* (R, S).

Remark 27. The analogous lemma for V-tame retractions appears in the book [dPW95, Proposition 9.3.20].

Proof. See Figure 2 for an illustration of the target situation.

- i) The composition $s \circ j_2$ is a homeomorphism by [dPW95, Proposition 9.3.6], and the proof that $r \circ i_2$ is a homeomorphism germ goes as in [dPW95, Proposition 9.3.19].
- ii) Form a retraction $(R, S) = ((r \circ i_2)^{-1} \circ r, (s \circ j_2)^{-1} \circ s)\colon F \to g$, and note that it has the same fibers as (r, s), hence must be E-tame.

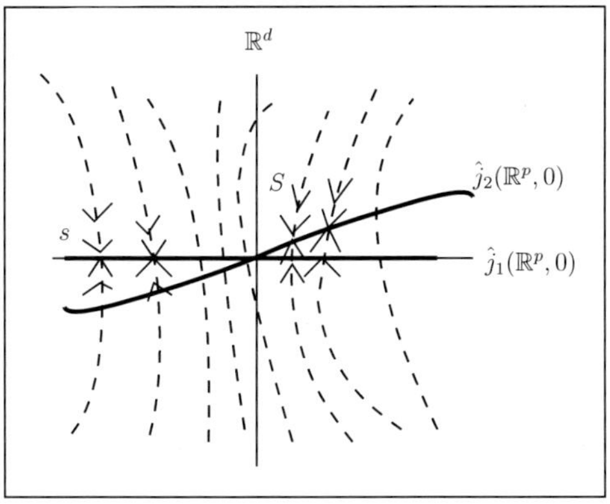

FIGURE 2. *Target situation:* The E-tame retraction $(r, s)\colon F \to f$ induces an E-tame retraction $(R, S)\colon F \to g$, whose fibers coincide with those of (r, s).

Note furthermore that if (r, s) is stratified smooth, then the homeomorphisms in i) are stratified diffeomorphisms, and the retractions (R, S) are stratified smooth as well. □

Proposition 28. *If $F\colon (N', x) \to (P', y)$ unfolds $f\colon (N, x) \to (P, y)$ stably, then any E-tame retraction $(r, s)\colon F \to f$ will leave any ST-invariant stratification of F invariant.*

Proof. Choose representatives $f\colon V \to W$ and $F\colon V \times U \to W \times U$, and suppose that $y \in S$ for an ST-invariant target stratum S. Now $\hat{F}_y$ and $\hat{F}_{s(y)}$ are $\mathscr{A}_0$-equivalent via the homeomorphisms $((r, \mathrm{pr}_U), (s, \mathrm{pr}_U))$. Hence $s(y) \in S$, and thus s leaves ST-invariant target strata invariant. But then r leaves the corresponding source strata invariant as well. □

2.3. **Combining retractions.** We are going to need some tools for combining E-tame retractions. First we see how E-tame retractions may be combined to yield new E-tame retractions in products of maps, such as in a multigerm situation.

Lemma 29. *Suppose that*

$$f = \bigsqcup_{i=1}^{s} \sigma_i \circ (f_i \times \mathrm{id}_{\prod_{j\neq i} \mathbb{R}^{p_j}})\colon \bigsqcup_{i=1}^{s} \mathbb{R}^{n_i} \times \prod_{j\neq i} \mathbb{R}^{p_j} \to \prod_{j=1}^{s} \mathbb{R}^{p_j}$$

is a multigerm, where σ_i is the permutation $\mathbb{R}^{p_i} \times \prod_{j\neq i} \mathbb{R}^{p_j} \to \prod_{j=1}^{s} \mathbb{R}^{p_j}$, and that

$$F = \bigsqcup_{i=1}^{s} \tilde{\sigma}_i \circ (F_i \times \mathrm{id}_{\prod_{j\neq i} \mathbb{R}^{p_j+d_j}})\colon \bigsqcup_{i=1}^{s} \mathbb{R}^{n_i+d_i} \times \prod_{j\neq i} \mathbb{R}^{p_j+d_j} \to \prod_{j=1}^{s} \mathbb{R}^{p_j+d_j}$$

is its unfolding, where F_i unfolds f_i for each $i \in \{1, \dots, s\}$, and $\tilde{\sigma}_i$ is the permutation $\mathbb{R}^{p_i+d_i} \times \prod_{j\neq i} \mathbb{R}^{p_j+d_j} \to \prod_{j=1}^{s} \mathbb{R}^{p_j+d_j}$. Assume furthermore that the $(r_i, s_i)\colon F_i \to f_i$ are E-tame retractions. Then

$$(r_i \times \prod_{j\neq i} s_j, \prod_{j=1}^{s} s_j) \tag{30}$$

is an E-tame retraction $F \to f$. If the (r_i, s_i) are smooth, then so is (30).

Proof. Trivial. □

We do not generally know how to glue local E-tame retractions together, but let us consider a situation where one of the retractions is smooth, and where we want to glue along a "line", for instance by using a distance function. More precisely, suppose that we are retracting onto source and target submanifolds N_0 and P_0, which are open subsets of $N \times \mathbb{R}$ and $P \times \mathbb{R}$ for smooth manifolds N and P, respectively. Here, we shall "glue" two retractions together while moving along the $\mathbb{R}$-component.

Let us assume (as we may, up to a reparametrization of $\mathbb{R}$) that

$$N \times [0, 3] \subset N_0,$$
$$P \times [0, 3] \subset P_0.$$

We assume, furthermore, that we are retracting onto a smooth map

$$f\colon N_0 \to P_0$$

where $f(N \times \{t\}) \subset P \times \{t\}$ for all t. We are retracting from the d-parameter unfolding

$$F\colon \tilde{N} \to \tilde{P}$$

of f, where $\tilde{N}$ and $\tilde{P}$ are open neighborhoods of $N_0 \times \{0\}$ and $P_0 \times \{0\}$ in $(N \times \mathbb{R}) \times \mathbb{R}^d$ and $(P \times \mathbb{R}) \times \mathbb{R}^d$, respectively, and F is $\mathbb{R}^d$-level preserving.

The retractions which we want to combine along $\mathbb{R}$ are given by

$$(r_i, s_i)\colon F \to f, \quad i = 1, 2, \tag{31}$$

where (r_1, s_1) is smooth and (r_2, s_2) is E-tame.

The first thing to note here is that by choosing suitable local coordinates at N_0 and P_0 and possibly shrinking $\tilde{N}$ and $\tilde{P}$, we may assume that (r_1, s_1) is a projection and $F = f \times \mathrm{id}_{\mathbb{R}^d}$. To see this, define maps

$$\Phi_N\colon \tilde{N} \to N_0 \times \mathbb{R}^d, \quad \Phi_N(y, t, u) = (s_1(y, t, u), u),$$
$$\Phi_P\colon \tilde{P} \to P_0 \times \mathbb{R}^d, \quad \Phi_P(x, t, u) = (r_1(x, t, u), u).$$

By shrinking $\tilde{N}$ and $\tilde{P}$ we may assume that Φ_N and Φ_P are diffeomorphisms, by Lemma 24. Furthermore, $s_1 \circ \Phi_N^{-1}$ and $r_1 \circ \Phi_P^{-1}$ agree with the projections $N_0 \times \mathbb{R}^d \to N_0$ and $P_0 \times \mathbb{R}^d \to P_0$ on $\tilde{N}$ and $\tilde{P}$, respectively. The map $s_2 \circ \Phi_N^{-1}\colon \Phi_N(\tilde{N}) \to N_0$ is E-tame, since Φ_N is a diffeomorphism.

We need, however, to see how the maps f and F behave with respect to this change of coordinates. Since

$$\begin{aligned}(\Phi_P \circ F)(x,t,u) &= (s_1(F(x,t,u)),u)\\ &= (f(r_1(x,t,u)),u)\\ &= (f \times \mathrm{id}_{\mathbb{R}^d})(\Phi_N(x,t,u)),\end{aligned}$$

we see that the diagram

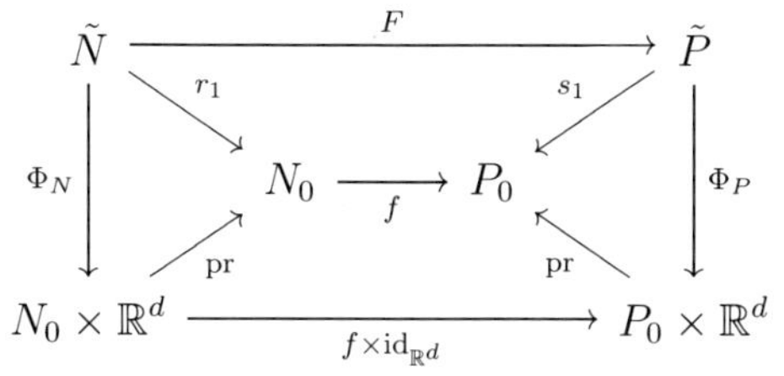

commutes, and in these new coordinates (r_1, s_1) are the projections, and F is of the form $f \times \mathrm{id}_{\mathbb{R}^d}$.

Hence we can assume from the start that F is just $f \times \mathrm{id}_{\mathbb{R}^d}$ and that

$$(r_1, s_1) = (\mathrm{pr}_{N_0}, \mathrm{pr}_{P_0}).$$

In the following lemma we shall prove that we can, indeed, combine the smooth projection with an E-tame retraction along $\mathbb{R}$:

Lemma 32. *Suppose that we have defined retractions* $(r_i, s_i)\colon F \to f$ *as described above, where* $(r_1, s_1) = (\mathrm{pr}, \mathrm{pr})$ *and* (r_2, s_2) *is E-tame, and where* $F = f \times \mathrm{id}_{\mathbb{R}^d}$. *Then, allowing for shrinking* $\tilde{N}$ *and* $\tilde{P}$, *we can find an E-tame retraction* $(R, S)\colon F \to f$ *such that*

$$\begin{aligned}&R = r_1 \text{ in } R^{-1}(N_0 \cap N \times [0,1]), \quad && S = s_1 \text{ in } S^{-1}(P_0 \cap P \times [0,1]),\\ &R = r_2 \text{ in } R^{-1}(N_0 \cap N \times [2,3]), \quad && S = s_2 \text{ in } S^{-1}(P_0 \cap P \times [2,3]).\end{aligned}$$

If (r_2, s_2) *is smooth, then so is* (R, S).

What seems like (but, as we have seen, is not) a more general result, is that we can combine in the same way the retractions (r_i, s_i) *given in (31).*

Remark 33. The non-relative case where we combine two retractions $s_1, s_2\colon \tilde{P} \to P_0$ with s_1 smooth and s_2 E-tame, follows from the general case by setting $f = \mathrm{id}_{P_0}$ and $F = \mathrm{id}_{\tilde{P}}$.

Proof. Pick a smooth map $\alpha\colon [1,2] \to [0,1]$ such that $\alpha(1) = 0, \alpha([2-\epsilon, 2]) = 1$ for some $\epsilon > 0$, $\alpha^{(n)}(1) = 0$ for all $n \in \mathbb{N}$, where (n) denotes the nth derivative, and $\alpha|(1,2] > 0$.

The idea is to stretch out the fibers of (r_2, s_2) as we approach $t = 1$ in the interval $]1,2[$ such that the fibers approach those of (r_1, s_1); we define $S\colon \tilde{P} \to P_0$ and

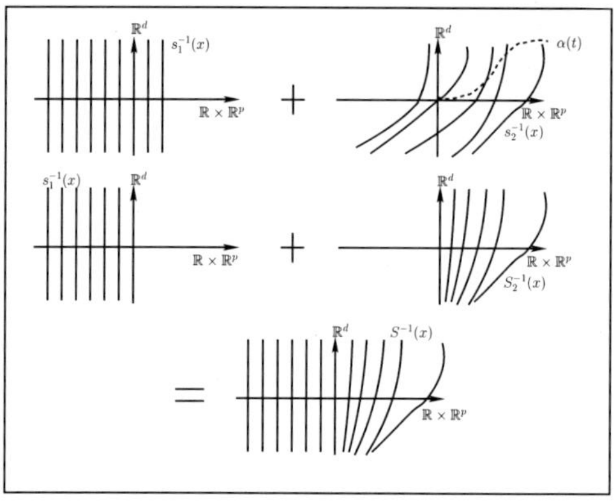

FIGURE 3. Combining a smooth retraction with an E-tame retraction along $[0,3]$.

$R\colon \tilde{N}\to N_0$ by setting

$$S(y,t,u) = \begin{cases} (y,t) & \text{if } t\in[0,1], \\ s_2(y,t,u) & \text{if } s_2(y,t,u)\in P\times[2,\infty), \\ s_2(y,t,\alpha(t)u) & \text{if } s_2(y,t,\alpha(t)u)\in P\times[1,2], \text{ and } t\in[1,2], \end{cases}$$
$$R(x,t,u) = \begin{cases} (x,t) & \text{if } t\in[0,1], \\ r_2(x,t,u) & \text{if } r_2(x,t,u)\in N\times[2,\infty), \\ r_2(x,t,\alpha(t)u) & \text{if } r_2(x,t,\alpha(t)u)\in N\times[1,2], \text{and } t\in[1,2]. \end{cases}$$

See Figure 3 for illustration.

If we define maps

$$\Omega_N\colon N\times\mathbb{R}\times\mathbb{R}^d\to N\times\mathbb{R}\times\mathbb{R}^d, (x,t,u)\mapsto(x,t,\alpha(t)u)\,,$$
$$\Omega_P\colon P\times\mathbb{R}\times\mathbb{R}^d\to P\times\mathbb{R}\times\mathbb{R}^d, (y,t,u)\mapsto(y,t,\alpha(t)u)\,,$$

(where we extend α to $\mathbb{R}$ continuously by setting $\alpha|]-\infty,1[\equiv 0$ and $\alpha|2,\infty[\equiv 1$), then we see that $R=r_2\circ\Omega_N$ and $S=s_2\circ\Omega_P$, so in particular, S and R are continuous, and we also easily see that (R,S) is a retraction $F=f\times\mathrm{id}_{\mathbb{R}^d}\to f$.

2.3.1. *S is E-tame at $P\times\{1\}\times\mathbb{R}^d$.* We shall prove that S is E-tame at $P\times\{1\}\times\mathbb{R}^d$, which is a bit tricky. We prove that if $(y,1,u)\in P\times\{1\}\times\mathbb{R}^d$ and $(y_n,t_n,u_n)\in P\times(1,2]\times\mathbb{R}^d$ such that $(y_n,t_n,u_n)\to(y,1,u)$, then

$$T_{(y_n,t_n,u_n)}S^{-1}(S(y_n,t_n,u_n))\to\mathbb{R}^d=\{y\}\times\{1\}\times\mathbb{R}^d.$$

In order to show this it is enough to show that for any $l\in\{1,\dots,d\}$, there exists

$$v_n\in T_{(y_n,t_n,u_n)}S^{-1}(S(y_n,t_n,u_n))$$

such that $\lim_{n\to\infty}v_n=e_l:=(0,\dots,1\dots,0)\in\mathbb{R}^d$, where e_l has 1 in the lth component. Fix $l\in\{1,\dots,d\}$; we construct such a sequence (v_n).

We define $\tilde{u}_n = \alpha(t_n)u_n$ and consider the sequence $(y_n, t_n, \tilde{u}_n)$, along with the sequence of s_2-fibers containing $(y_n, t_n, \tilde{u}_n)$. Clearly, $(y_n, t_n, \tilde{u}_n) \to (y, 1, 0)$.

There is a sequence of maps

$$\beta_n \colon W \to P \times [0,3] \times \mathbb{R}^d$$

for a neighborhood W of 0 in $\mathbb{R}^d$ such that (shrinking $\tilde{P}$) β_n is a diffeomorphism onto $s_2^{-1}(s_2(x_n, t_n, \tilde{u}_n))$ for each n, and such that $\mathrm{pr}_{\mathbb{R}^d} \circ \beta_n = \mathrm{id}_W$. This holds because the fibers form a foliation transverse to $P \times [0,3]$.

We write

$$\beta_n = (\beta_{nP}, \beta_{n[0,3]}, \beta_{n\mathbb{R}_1}, \ldots, \beta_{n\mathbb{R}_d}) = (\beta_{n1}, \beta_{n2}, \beta_{n31}, \ldots, \beta_{n3d});$$

now the second condition on β_n is equivalent to $\beta_{n3i} = \mathrm{pr}_{\mathbb{R}_i}$.

Define a path

$$\lambda_n \colon \mathbb{R} \to \mathbb{R}^d, \quad \lambda_n(t) = t \cdot e_l + \tilde{u}_n.$$

Then $\lambda_n'(0) = e_l \in T_{\tilde{u}_n}\mathbb{R}^d = \mathbb{R}^d$, and $(\lambda_n)_i'(0) = \delta_{il}$.

Consider the vectors

$$\tilde{v}_n = (\beta_n \circ \lambda_n)'(0) \in T_{(x_n, t_n, \tilde{u}_n)} s_2^{-1}(s_2(x_n, t_n, \tilde{u}_n)).$$

Then we have

$$\begin{aligned} \tilde{v}_n &= ((\beta_{n1} \circ \lambda_n)'(0), (\beta_{n2} \circ \lambda_n)'(0), D\lambda_n(0)) \\ &= ((\beta_{n1} \circ \lambda_n)'(0), (\beta_{n2} \circ \lambda_n)'(0), \underbrace{0, \ldots, 1, \ldots, 0}_{e_l}), \end{aligned}$$

and we know that s_2 is E-tame, meaning that its fibers form a $C^{0,1}$-foliation. Hence we have $\tilde{v}_n \to \tilde{v}$ for some $\tilde{v} \in T_{(y,1,0)} s_2^{-1}(s_2(y, 1, 0))$ such that $\mathrm{pr}_{\mathbb{R}^d}(\tilde{v}) = \lim \mathrm{pr}_{\mathbb{R}^d}(\tilde{v}_n) = \lim e_l = e_l$.

Define

$$\gamma_n \colon W \to P \times [0,3] \times \mathbb{R}^d,$$

by

$$\gamma_n(u) = (\beta_{n1}(u), \beta_{n2}(u), \frac{1}{\alpha(\beta_{n2}(u))}\beta_{n3}(u)) = (\beta_{n1}(u), \beta_{n2}(u), \frac{u}{\alpha(\beta_{n2}(u))}).$$

Now γ_n is a diffeomorphism onto $S^{-1}(S(x_n, t_n, u_n))$.

Define a sequence of vectors $(v_n) \in T_{(x_n, t_n, u_n)} S^{-1}(S(x_n, t_n, u_n))$ by setting

$$v_n = (\gamma_n \circ \lambda_n)'(0).$$

Clearly,

$$\begin{aligned} \mathrm{pr}_P(v_n) &= \mathrm{pr}_P \tilde{v}_n, \text{ and} \\ \mathrm{pr}_{[0,3]} v_n &= \mathrm{pr}_{[0,3]} \tilde{v}_n. \end{aligned}$$

We consider $v_{n3i} = \mathrm{pr}_{\mathbb{R}_i} v_n$:

$$\begin{aligned} v_{n3i} &= \tfrac{d}{ds}(\gamma_{n3i} \circ \lambda_n(s))\big|_{s=0} \\ &= \tfrac{d}{ds}\left(\frac{(\lambda_n(s))_i}{\alpha(\beta_{n2}(\lambda_n(s)))}\right)\Big|_{s=0} \\ &= \frac{\alpha(\beta_{n2}(\lambda_n(s)))\cdot(\lambda_n)_i'(s) - (\lambda_n(s))_i\cdot(\alpha'(\beta_{n2}(\lambda_n(s)))\cdot(\beta_{n2}\circ\lambda_n)'(s))}{\alpha(\beta_{n2}(\lambda_n(s)))^2}\Big|_{s=0} \\ &= \frac{\alpha(t_n)\cdot\delta_{il} - (\tilde{u})_i\cdot(\alpha'(t_n)\cdot\tilde{v}_{n2})}{\alpha(t_n)^2} \\ &= \frac{\delta_{il}}{\alpha(t_n)} - \frac{\alpha(t_n)(u_n)_i\cdot\alpha'(t_n)\cdot(\tilde{v}_n)_2}{\alpha(t_n)^2} \\ &= \frac{\delta_{il}}{\alpha(t_n)} - \frac{(u_n)_i\cdot\alpha'(t_n)\cdot(\tilde{v}_n)_2}{\alpha(t_n)}. \end{aligned}$$

We see that for $i \neq l$ we have

$$\frac{v_{n3l}}{v_{n3i}} = \frac{1 - (u_n)_l \cdot \alpha'(t_n) \cdot (\tilde{v}_n)_2}{(u_n)_i \cdot \alpha'(t_n) \cdot (\tilde{v}_n)_2} \to \infty,$$

because $u_n \to u$ and $(\tilde{v}_n)_2 \to \tilde{v}_2$, giving $(u_n)_i \cdot \alpha'(t_n) \cdot (\tilde{v}_n)_2 \to 0$.

Similarly,

$$\frac{v_{n3l}}{\|\mathrm{pr}_P(v_n)\|} \to \infty, \qquad \frac{v_{3nl}}{\|\mathrm{pr}_{[0,3]}(v_n)\|} \to \infty,$$

giving

$$\frac{v_n}{\|v_n\|} \to e_l,$$

which proves our claim.

2.3.2. *If s_2 is smooth, then so is S.* Smoothness is clear off $P \times \{1\} \times \mathbb{R}^d$, and as $(y, t, u) \to (y_0, 1_-, u_0)$. Hence it remains to show that the partial derivatives of order d are continuous in the limit $(y, t, u) \to (y_0, 1_+, u_0)$. Let us take a look at the total derivative of S at a point (y, t, u) where $t \in]1, 2]$.

Since $S(y, t, u) = s_2(y, t, \alpha(t)u)$, we get

$$DS(y, t, u) = Ds_2(\Psi(y, t, u)) \circ D\Psi(y, t, u),$$

where $\Psi\colon P \times (1, 2] \times \mathbb{R}^d \to P \times (1, 2] \times \mathbb{R}^d$ is given by $\Psi(y, t, u) = (y, t, \alpha(t)u)$. If s_2 is C^k-smooth ($k \geq 1$), then the map Ds_2 is C^{k-1}, and in matrix form we can write

$$Ds_2(y, t, u) = \begin{pmatrix} A_{nn}(y,t,u) & A_{nt}(y,t,u) & A_{nd}(y,t,u) \\ A_{tn}(y,t,u) & A_{tt}(y,t,u) & A_{td}(y,t,u) \end{pmatrix}$$

where, in particular, we write

$$A_{nd}(y, t, u) = \begin{pmatrix} a_{11}(y,t,u) & \ldots & a_{1d}(y,t,u) \\ \vdots & & \vdots \\ a_{n1}(y,t,u) & \ldots & a_{nd}(y,t,u) \end{pmatrix}$$

$$A_{td}(y, t, u) = \begin{pmatrix} b_1(y,t,u) & \ldots & b_d(y,t,u) \end{pmatrix}, \text{ and}$$

$$A_{nt}(y, t, u) = \begin{pmatrix} c_1(y,t,u) \\ \vdots \\ c_n(y,t,u) \end{pmatrix},$$

and we have

$$\begin{pmatrix} A_{nn} & A_{nt} \\ A_{tn} & A_{tt} \end{pmatrix} (y_0, t_0, 0) = I_{n+1},$$

since s_2 is a retraction. The map Ψ is C^∞ and has the total derivative

$$D\Psi(y,t,u) = \begin{pmatrix} I & & & & \\ & 1 & \alpha'(t)u_1 & \dots & \alpha'(t)u_d \\ & \alpha'(t)u_1 & \alpha(t) & & \\ & \vdots & & \ddots & \\ & \alpha'(t)u_d & & & \alpha(t) \end{pmatrix},$$

where

$$I = \begin{pmatrix} 1 & & \\ & \ddots & \\ & & 1 \end{pmatrix},$$

and $u = (u_1, \dots, u_d)$. It follows that for $(y,t,u) \in P \times]1,2] \times \mathbb{R}^d$, we get

$$Ds_2(\Psi(y,t,u)) \circ D\Psi(y,t,u)$$

$$= \begin{pmatrix} A_{nn} & B_{nt} & B_{nd} \\ A_{tn} & B_{tt} & B_{td} \end{pmatrix}, \tag{34}$$

where

$$B_{nt} = \begin{pmatrix} c_1 + \alpha'(t) \sum_{i=1}^d u_i a_{1i} \\ \vdots \\ c_n + \alpha'(t) \sum_{i=1}^d u_i a_{ni} \end{pmatrix},$$

$$B_{tt} = A_{tt} + \alpha'(t) \sum_{i=1}^d u_i b_i,$$

$$B_{nd} = \begin{pmatrix} \alpha'(t)u_1 c_1 + \alpha(t)a_{11} & \dots & \alpha'(t)u_d c_1 + \alpha(t)a_{1d} \\ \vdots & & \vdots \\ \alpha'(t)u_1 c_n + \alpha(t)a_{n1} & \dots & \alpha'(t)u_d c_n + \alpha(t)a_{nd} \end{pmatrix},$$

$$B_{td} = \begin{pmatrix} \alpha'(t)u_1 A_{tt} + \alpha(t)b_1 & \dots & \alpha'(t)u_d A_{tt} + \alpha(t)b_d \end{pmatrix},$$

evaluated at $(y, t, \alpha(t)u)$. This expression approaches

$$\begin{pmatrix} A_{nn}(y_0,1,0) & A_{nt}(y_0,1,0) & 0 \\ A_{tn}(y_0,1,0) & A_{tt}(y_0,1,0) & 0 \end{pmatrix} = \begin{pmatrix} I_{n+1} & 0 \end{pmatrix} = DS(y_0, 1, u_0)$$

as $(y,t,u) \to (y_0, 1, u_0)$, proving that DS is continuous, and thus that S is C^1 smooth. Looking at the entries of (34), which are just the first order partial derivatives of the component functions of S, it is easy to see that we may continue to differentiate each and every one of them arbitrarily many times more, and that all these derivatives will vanish at $(y_0, 1, u_0)$, since we know that $\alpha^{(n)}(1) = 0$ for all $n \in \mathbb{N}$ (we know that the higher order partial derivatives of entries of A_{nn}, A_{nt}, A_{tn}

and A_{tt} with respect to t and coordinates of y, must vanish at $(y_0, 1, u_0)$ because s_2 is a retraction).

Remark 35. There are other ways to combine C^k-smooth retractions, e.g. by combining their corresponding vector fields using a partition of unity. But the current approach will allow us to combine a smooth retraction with a stratified smooth, E-tame retraction to get an E-tame, stratified smooth retraction.

□

2.4. **Weighted homogeneous maps and E-tame retractions.** A map $f\colon \mathbb{R}^n \to \mathbb{R}^p$ is said to be *weighted homogeneous* if it is equivariant with respect to $\mathbb{R}^+$-actions ($\mathbb{R}^+ = (]0, \infty[, \cdot)$) on source and target of the form

$$t \cdot (x_1, \dots, x_n) = (t^{a_1}x_1, \dots, t^{a_n}x_n), \qquad t \cdot (y_1, \dots y_p) = (t^{b_1}y_1, \dots t^{b_p}y_p),$$

that is, if $f(t \cdot x) = t \cdot f(x)$ for all $t \in \mathbb{R}^+$ and all $x \in \mathbb{R}^n$. Here the a_i are called *source weights* and the b_i are called *target weights.*

We define weighted distance functions $\sigma\colon \mathbb{R}^n \to \mathbb{R}$ and $\rho\colon \mathbb{R}^p \to \mathbb{R}$ on source and target as follows:

$$\sigma(x) = \sum \left\{ x_i^{A/a_i} | 1 \le i \le n \text{ s.t. } a_i > 0, A = \operatorname{lcm}\{2a_i | 1 \le i \le n,\ a_i > 0\} \right\},$$
$$\rho(y) = \sum \left\{ y_i^{B/b_i} | 1 \le i \le p \text{ s.t. } b_i > 0, B = \operatorname{lcm}\{2b_i | 1 \le i \le p,\ b_i > 0\} \right\}.$$

The functions σ and ρ measure the distance from the non-positively weighted subspaces of source and target, respectively.

In the case of weighted homogeneous maps, we can sometimes reduce the construction of E-tame retractions to a problem on lower-dimensional slices defined by the distance functions. More precisely:

Lemma 36. *Let $F\colon N \times U \to P \times U$ be a weighted homogeneous, U-level preserving unfolding of $F^+\colon N \to P$ which has FST, and assume that the weights on N and P are positive, and that the weights on U are non-positive. Here U is a real vector space $\mathbb{R}^d$, and we decompose $U = U_0 \oplus U_-$, where U_0 has zero weight, and U_- has negative weights.*

Denote by F_ϵ the restriction $F|\colon F^{-1}\rho^{-1}(\epsilon) \to \rho^{-1}(\epsilon)$, and similarly for the positively weighted part F^+ of F.

Suppose that $(\hat{r}, \hat{s})$ is a germ at

$$(N \times \{0_U\}, P \times \{0_U\}) \cap (F^{-1}\rho^{-1}(\epsilon), \rho^{-1}(\epsilon))$$

of E-tame retraction to the inclusion $F_\epsilon^+ \to F_\epsilon$.

Then there exist $\mathbb{R}^+$-invariant neighborhoods W_N of $(N \times \{0_U\}) \cup (\{0_N\} \times U_-)$ in $N \times U$, and W_P of $(P \times \{0_U\}) \cup (\{0_P\} \times U_-)$ in $P \times U$, with $F(W_N) \subset W_P$, and an $\mathbb{R}^+$-equivariant E-tame retraction

$$(R, S)\colon \{F|\colon W_N \to W_P\} \to F^+$$

to the inclusion. Here N and P are identified with their images $N \times \{0\}$ and $P \times \{0\}$ in $N \times U$, $P \times U$, respectively.

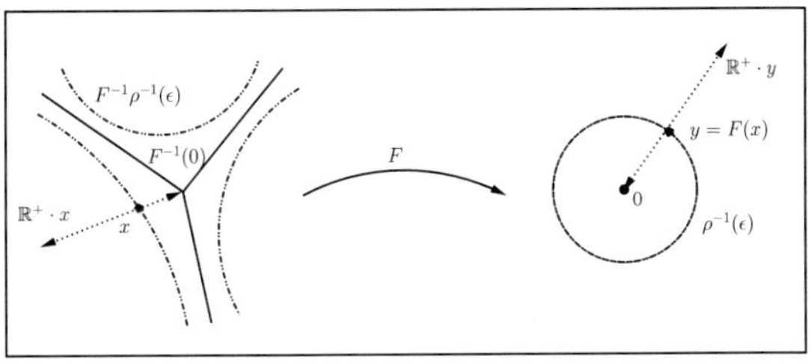

FIGURE 4. We extend the retraction (r_ϵ, s_ϵ) on the slices $(F^{-1}\rho^{-1}(\epsilon), \rho^{-1}(\epsilon))$ through the $\mathbb{R}^+$-action.

If the retraction $(\hat{r}, \hat{s})$ is smooth, then the retraction (R, S) will be stratified smooth with respect to the stratification

$$\left(\left\{0 \times U, F^{-1}\rho^{-1}(0) \setminus (0 \times U), N \times U \setminus F^{-1}\rho^{-1}(0)\right\}, \{0 \times U, P \setminus 0 \times U\}\right).$$

This is an E-tame version of lemmas appearing in the book by du Plessis and Wall [dPW95, Lemmas 9.6.3 and 9.6.4].

Remark 37. We have chosen to think of weighted homogeneous maps as $\mathbb{R}^+$-equivariant rather than $\mathbb{R}^*$-equivariant, as is done in the book [dPW95]. The reason for this choice is that $\mathbb{R}^+$ acts freely on $(P \setminus \{0\}) \times U$ with $\rho^{-1}(\epsilon)$ as a global slice, while the action of $\mathbb{R}^*$ is not free. Given any point $(x, u) \in (P \setminus \{0\}) \times U$ where all the coordinates of non-even weight are zero, the isotropy subgroup $\mathbb{R}^*_{(x,u)}$ is $\{\pm 1\}$. The level sets $\rho^{-1}(\epsilon)$ are $\{\pm 1\}$-invariant, and in order to get an $\mathbb{R}^*$-invariant retraction (R, S) in Lemma 36, as is stated in the book [dPW95, Lemma 9.6.4], we actually need (at least) the $(\hat{r}, \hat{s})$ to be $\{\pm 1\}$-equivariant. For most purposes this does not make any difference, but for our constructions of E-tame retractions in Chapter 4, the distinction will be important.

We denote by n and p the dimensions of N and P, respectively.

Proof. First of all, fix E-tame representatives (r, s) of the germ $(\hat{r}, \hat{s})$.

We start out by extending the target retraction s, and set $W_P = \mathbb{R}^+ \cdot s(s) \cup (\{0_P\} \times U)$. (Recall that $s(s)$ denotes the source of the retraction s, and that $t(s)$ would similarly denote the target of the same retraction.) Now W_P is a neighborhood of $(P \times \{0_U\}) \cup (\{0_P\} \times U_-)$, although it is not necessarily open. Let us agree that every time we shrink W_P, we shrink $s(s)$ correspondingly, so that the defining formula of W_P in terms of $s(s)$ still holds.

Since s is E-tame, the fibers of s define a $C^{0,1}$-foliation $\mathscr{F}_s$ on $s(s)$, whose leaves meet $P \times \{0\}$ transversely. We recall that constructing the target part of an E-tame retraction $F \to F^+$ is equivalent to finding a $C^{0,1}$-foliation near $t(F^+)$ in $t(F)$ of dimension d, which is transverse to $P \times \{0\}$.

The smooth, free $\mathbb{R}^+$-action on $(P \setminus \{0\}) \times U$ induces a canonical diffeomorphism

$$(P \setminus \{0\}) \times U = \mathbb{R}^+ \cdot \rho^{-1}(\epsilon) \cong \mathbb{R}^+ \times \rho^{-1}(\epsilon),$$

which allows us to define a foliation $\mathbb{R}^+ \times \mathscr{F}_s$ on $\mathbb{R}^+ \cdot s(s)$. By including the set $\{0_P\} \times U$ as a leaf, we obtain a partition $\mathscr{F}$ of W_P; we claim that this is a $C^{0,1}$-foliation.

Lemma 38. *Suppose that $P \times U$ has a weighted $\mathbb{R}^+$-action where the weights on P are positive, and the weights on U are non-positive, writing $U = U_0 \times U_-$ for the split into zero- and negatively weighted subspaces. Assume that $\dim(P) \geq 1$. Suppose that S is a [smooth] slice for the free $\mathbb{R}^+$-action on $(P \setminus \{0\}) \times U$, such that $S \cap (P \times \{0\})$ is compact. Suppose that a d-dimensional $C^{0,1}$-foliation $\mathscr{F}_s$ is defined on S, which is transverse to $P \times \{0\}$ in S. Extend it to a foliation $\mathscr{F}$ on $\mathbb{R}^+ \cdot S \cup \{0\} \times U$ by using the $\mathbb{R}^+$-action to reach $\mathbb{R}^+ \cdot S$ ($t\colon S \to t \cdot S$ is a homeomorphism [diffeomorphism]) and by taking $\{0\} \times U$ as the final leaf. Then $\mathscr{F}$ is a $C^{0,1}$-foliation on $\mathbb{R}^+ \cdot S \cup \{0\} \times U$, transverse to $P \times \{0\}$.*

Proof. Since the action of $\mathbb{R}^+$ is smooth, it is clear that $\mathscr{F} \cap (P \setminus \{0\} \times U)$ is a $C^{0,1}$-foliation. It remains to show that if

$$(y_m, u_m, u'_m) \in ((P \setminus \{0\}) \times U_0 \times U_-) \cap \mathbb{R}^+ \cdot S$$

such that $(y_m, u_m, u'_m) \to (0, u, u')$, then

$$T_{(y_m, u_m, u'_m)}\mathscr{F} \to T_{(0,u,u')}\mathscr{F} = \mathbb{R}^d.$$

In order to show this it is enough to show that for all $l \in \{1, \ldots, d\}$ there exists

$$v_m \in T_{(y_m, u_m, u'_m)}\mathscr{F} \ \forall \ m \in \mathbb{N},$$

such that

$$\lim v_m = e_l \in \mathbb{R}^d.$$

Fix l; we find a suitable sequence (v_m).

For each m there exists $t_m \in \mathbb{R}^+$ such that

$$t_m \cdot (y_m, u_m, u'_m) \in S$$

and there exists a map-germ

$$\beta_m\colon (\mathbb{R}^d, 0) \to (S, t_m \cdot (y_m, u_m, u'_m)),$$

which is a diffeomorphism onto the leaf of $\mathscr{F}_s$ containing $t_m \cdot (y_m, u_m, u'_m)$.

Now, the map-germ

$$t_m^{-1} \circ \beta_m\colon (\mathbb{R}^d, 0) \to (t_m^{-1} \cdot S, (y_m, u_m, u'_m))$$

takes $\mathbb{R}^d$ diffeomorphically to the leaf of $\mathscr{F}$ containing (y_m, u_m, u'_m). We write

$$\beta_m = (\beta_{m1}, \ldots, \beta_{mn}, \beta_{m(n+1)}, \ldots, \beta_{mk}, \beta_{m(k+1)}, \ldots, \beta_{mq}),$$

where $q = n + d$ and $k - n = \dim U_0$. Note that we may assume

$$(\beta_{m(n+1)}, \ldots, \beta_{mq}) = \mathrm{id}_{\mathbb{R}^d} \tag{39}$$

by choosing β_m suitably. If the action of $\mathbb{R}^+$ on $P \times U$ is given by

$$\begin{aligned} t \cdot (y_1, \ldots, y_n, u_{n+1}, \ldots, u_k, u'_{k+1}, \ldots, u'_q) \\ = (t^{b_1} y_1, \ldots, t^{b_n} y_n, u_{n+1}, \ldots, u_k, t^{b_{k+1}} u'_{k+1}, \ldots, t^{b_q} u'_q), \end{aligned}$$

where

$$b_i \begin{cases} > 0 & \text{if } i \leq n \\ < 0 & \text{if } i > k \end{cases}$$

then

$$t_m^{-1} \circ \beta_m$$

equals

$$((t_m^{-1})^{b_1}\beta_{m1}, \ldots, (t_m^{-1})^{b_n}\beta_{mn}, \beta_{m(n+1)}, \ldots, \beta_{mk}, \\ (t_m^{-1})^{b_{k+1}}\beta_{m(k+1)} \ldots (t_m^{-1})^{b_{mq}}\beta_{mq}).$$

Since $y_m \to 0$ we must have $t_m \to \infty$, giving $t_m^{-1} \to 0$.

Write $\lambda \colon \mathbb{R} \to \mathbb{R}^d$, $t \mapsto t \cdot e_l$, giving $\lambda'(0) = e_l \in \mathbb{R}^d$. Let

$$\tilde{v}_m = (\beta_m \circ \lambda)'(0) \in T_{t_m \cdot (y_m, u_m, u'_m)} \mathscr{F}_s.$$

Since $S \cap (P \times \{0\})$ is compact, there exists a subsequence (m_k) of (m) such that $t_{m_k} \cdot y_{m_k} \to \tilde{y} \in S \cap (P \times \{0\})$, so $t_m \cdot (y_m, u_m, u'_m) \to (\tilde{y}, u, 0)$. Then since $\mathscr{F}_s$ is a $C^{0,1}$-foliation, we have $\tilde{v}_{m_k} \to \tilde{v}$ for some $\tilde{v} \in T_{(\tilde{y},u,0)}\mathscr{F}_s$.

In particular, the set of $\tilde{v}_{m_k}$ is bounded, and then, since the β_m can be assumed to be restrictions of a $C^{0,1}$-differentiable map-germ $\beta \colon ((P \cap S) \times \mathbb{R}^d, (P \cap S) \times 0) \to S$, the set consisting of *all* the $\tilde{v}_m$ is bounded as well.

Define

$$v_m = \frac{(t_m^{-1} \circ \beta_m \circ \lambda)'(0)}{\| (t_m^{-1} \circ \beta_m \circ \lambda)'(0) \|},$$

then, calculating, we see that

$$\begin{aligned} v_m &= \frac{\frac{d}{ds}\begin{pmatrix} (t_m^{-1})^{b_1} \cdot (\beta_{m1} \circ \lambda)(s), \ldots, (t_m^{-1})^{b_n}(\beta_{mn}) \circ \lambda)(s), \\ (\beta_{m(n+1)} \circ \lambda)(s), \ldots, (\beta_{mk} \circ \lambda)(s), \\ (t_m^{-1})^{b_{k+1}}(\beta_{m(k+1)} \circ \lambda)(s), \ldots, (t_m^{-1})^{b_q}(\beta_{mq} \circ \lambda)(s) \end{pmatrix}\Bigg|_{s=0}}{\| - - - \|} \\ &= \frac{\begin{pmatrix} (t_m^{-1})^{b_1} \cdot (\beta_{m1} \circ \lambda)'(0), \ldots, (t_m^{-1})^{b_n}(\beta_{mn} \circ \lambda)'(0), \\ (\beta_{m(n+1)} \circ \lambda)'(0), \ldots (\beta_{mk} \circ \lambda)'(0), \\ (t_m^{-1})^{b_{k+1}} v_{m(k+1)}, \ldots, (t_m^{-1})^{b_q}(\beta_{mq} \circ \lambda)'(0) \end{pmatrix}}{\| - - - \|} \\ &= \frac{((t_m^{-1})^{b_1}\tilde{v}_{m1}, \ldots, (t_m^{-1})^{b_n}\tilde{v}_{mn}, \tilde{v}_{m(n+1)}, \ldots, \tilde{v}_{mk}, (t_m^{-1})^{b_{k+1}}\tilde{v}_{m(k+1)}, \ldots, (t_m^{-1})^{b_q}\tilde{v}_{mq})}{\| - - - \|} \\ &\to e_l \end{aligned}$$

since $(t_m^{-1})^{b_i}\tilde{v}_{mi} \to 0$ as $m \to \infty$ for $i \leq n$, and, by (39),

$$(\tilde{v}_{m(n+1)}, \ldots, \tilde{v}_{mk}, (t_m^{-1})^{b_{k+1}}\tilde{v}_{m(k+1)}, \ldots, (t_m^{-1})^{b_q}\tilde{v}_{mq}) \to \begin{cases} e_l & \text{if } e_l \in U_0 \\ \infty \cdot e_l & \text{if } e_l \in U_-. \end{cases}$$

Then (v_m) is the wanted sequence, and Lemma 38 holds. □

Now the construction of the target retraction $S \colon W_P \to W_P \cap P \times \{0\}$ is trivial; just follow the leaves of the foliation.

2.4.1. *Extension of source retraction.* In the source, we can just follow the proof by du Plessis and Wall [dPW95, Lemma 9.6.4], in order to get a continuous retraction R that forms an E-tame retraction $F \to F^+$ together with S.

However, if the level set retraction (r, s) is smooth, then we can actually construct a retraction (R, S) which is stratified smooth with respect to the stratification described in the statement of the lemma.

When s is smooth, it is clear that the E-tame retraction S, which we constructed above, is stratified smooth; we prove the claim for R.

As described on p. 25, we can fix smooth vector fields

$$\tilde{\eta}_i \colon s(s) \to Ts(s), \quad \tilde{\xi}_i \colon s(r) \to Ts(r),$$

where the $\tilde{\eta}_i$ lift the standard vector fields $\frac{\partial}{\partial u_i}$ on $U \subset \mathbb{R}^d$ over pr and the $\tilde{\xi}_i$ lift the $\tilde{\eta}_i$ over F, such that the $\tilde{\eta}_i$ and $\tilde{\xi}_i$ induce s and r, and are tangent to the fibers.

Then the foliation $\mathscr{F}$ is spanned by the naturally extended vector fields η_i on $\mathbb{R}^+ \cdot \rho^{-1}(\epsilon)$ and by $0 \times \frac{\partial}{\partial u_i}$ on $\{0\} \times U$, and we may assume that the η_i lift the $\frac{\partial}{\partial u_i}$ over pr. We aim to also extend the $\tilde{\xi}_i$ to vector fields ξ_i in a neighborhood of $(N \times \{0_U\}) \cup (\{0_N\} \times U_-)$ using the $\mathbb{R}^+$-action, in such a way that the ξ_i lift the η_i over F, and such that the ξ_i are continuous, stratified smooth and integrable.

By [dPW95, p. 392] there exists a neighborhood W of $F^{-1}(0,0) \cap \sigma^{-1}(\delta)$ (for sufficiently small $\delta > 0$) in $N \times U$ where $(\sigma, F) \colon N \times U \to \mathbb{R} \times P \times U$ is a proper submersion, i.e. a bundle projection by the Ehresmann fibration lemma, and the local picture has the form

$$\begin{array}{ccc} X \times \mathbb{R} \times P & \xrightarrow{(\sigma|N, F^+) = \mathrm{pr}} & \mathbb{R} \times P \\ \downarrow & & \downarrow \\ X \times \mathbb{R} \times P \times U & \xrightarrow[(\sigma, F) = \mathrm{pr}]{} & \mathbb{R} \times P \times U \end{array}$$

Choose $\epsilon > 0$ small enough that the set

$$A_1 = (\rho \circ F)^{-1}([0, \epsilon]) \cap \sigma^{-1}(\delta) \cap W_N,$$

is contained in W. Note that A_1 is a smooth slice for the $\mathbb{R}^+$-action in $N \times U$.

Any vector field ξ on $X \times \mathbb{R} \times P \times U$ can be written as a product

$$(\xi^{(1)}, \xi^{(2)}, \xi^{(3)}, \xi^{(4)}) \colon X \times \mathbb{R} \times P \times U \to TX \oplus T\mathbb{R} \oplus TP \oplus TU = T(X \times \mathbb{R} \times P \times U);$$

given a vector field $\bar{\eta}$ on $\mathbb{R} \times P \times U$ it can be written as a product

$$\bar{\eta} = (\bar{\eta}^{(2)}, \bar{\eta}^{(3)}, \bar{\eta}^{(4)}) \colon \mathbb{R} \times P \times U \to T\mathbb{R} \oplus TP \oplus TU = T(\mathbb{R} \times P \times U);$$

and given a vector field η on $P \times U$ it can be written as a product

$$\eta = (\eta^{(3)}, \eta^{(4)}) \colon P \times U \to TP \oplus TU = T(P \times U).$$

Now, ξ lifts $\bar{\eta}$ over (σ, F) in W if and only if $\xi^{(i)} = \bar{\eta}^{(i)}$ in W for $i = 2, 3, 4$ and ξ lifts η over F in W if and only if $\xi^{(i)} = \eta^{(i)}$ for $i = 3, 4$ in W.

Thus, given a vector field η on $P \times U$, define $\bar{\eta} = (0, \eta)$ on $\mathbb{R} \times (P \times U)$ and lift it to $\xi = (0, \bar{\eta}) = (0, 0, \bar{\eta})$ on $X \times \mathbb{R} \times P \times U$. If η was stratified smooth and continuous on $(P \times U, (0,0))$ then ξ will be stratified smooth and continuous on $(N \times U, (0,0))$.

Recall that we have stratified smooth, continuous vector fields $\eta_1, \ldots, \eta_d$ on $W_P \subset P \times U$ lifting $\frac{\partial}{\partial u_1}, \ldots, \frac{\partial}{\partial u_d}$ from U as above, which are tangent to the level sets $\rho^{-1}(\epsilon)$. By the preceding discussion we can lift the η_i to stratified smooth, continuous vector fields $\xi_1, \ldots, \xi_d$ on A_1.

Note that the ξ_i will be tangent both to the level sets $\sigma^{-1}(\delta')$ and to $F^{-1}\rho^{-1}(\epsilon')$ for δ' near δ and ϵ' near ϵ.

We extend the vector fields to stratified smooth, continuous vector fields on $\mathbb{R}^+ \cdot A_1$ by setting $\xi_i(t \cdot x) = h \cdot (t_*(\xi_i(x)))$, where $h \colon P \times U \to \mathbb{R}$ is the normalization factor which ensures that the ξ_i lift the $\frac{\partial}{\partial u_i}$ over the projection. Note that within W, the X- and $\mathbb{R}$- coordinates of ξ_i are 0, and note that off $F^{-1}\rho^{-1}(0)$, the ξ_i are smooth.

Next we consider the set

$$A_2 = (\rho \circ F)^{-1}(\epsilon) \cap \sigma^{-1}([0, \delta]) \cap W_N,$$

which is a smooth slice for the $\mathbb{R}^+$-action, and we also note that $A_1 \cup A_2$ is a global topological slice for $W_N \setminus \{0_N\} \times U$.

In A_2 we have a retraction r defined through smooth vector fields $\tilde{\xi}_i$ from the very beginning. The problem is just that these vector fields $\tilde{\xi}_i$ do not necessarily agree with the ξ_i which we have defined on A_1 in the common domain $A_1 \cap A_2 = \sigma^{-1}(\delta) \cap (\rho \circ F)^{-1}(\epsilon)$, and they certainly need not combine to give a stratified smooth vector field.

We deform the $\tilde{\xi}_i$ near $\sigma^{-1}(\delta)$ in order to get them to agree with the previously defined ξ_i on A_1.

Close to $\sigma^{-1}(\delta)$, we are still within the product representation $X \times \mathbb{R} \times P \times U$, and here $\tilde{\xi}_i$ is of the form

$$\tilde{\xi}_i = (\tilde{\xi}_i^{(1)}, \tilde{\xi}_i^{(2)}, \tilde{\xi}_i^{(3)}, \tilde{\xi}_i^{(4)}),$$

and we want to eliminate the first two components as we approach $\sigma^{-1}(\delta)$. We can obtain such a situation through multiplying the vector field coordinates in these directions by a bump function which is zero close to $\sigma^{-1}(\delta)$.

The problem with this approach is to show that our vector fields are still integrable after our treatment.

Let us look at the technicalities.

In W the vector field $\tilde{\xi}_i$ will be of the form

$$\tilde{\xi}_i = (\tilde{\xi}_i^{(1)}, \tilde{\xi}_i^{(2)}, \tilde{\xi}_i^{(3)}, \tilde{\xi}_i^{(4)}) \colon$$
$$(X \times \mathbb{R} \times P \times U, X \times \{\delta\} \times \{0\} \times \{0\}) \to TX \times T\mathbb{R} \times TP \times TU,$$

and we pick a C^∞-function

$$\chi \colon \mathbb{R} \to \mathbb{R}$$

such that χ is 0 near δ and χ is 1 on $\mathbb{R} \setminus [\delta - \delta', \delta + \delta']$, for some small $\delta' > 0$.

We define new vector fields ξ_i by setting

$$\xi_i = (\chi \cdot \tilde{\xi}_i^{(1)}, \chi \cdot \tilde{\xi}_i^{(2)}, \tilde{\xi}_i^{(3)}, \tilde{\xi}_i^{(4)}),$$

and we extend to the $\mathbb{R}^+$-saturation of A_2 by setting $\xi_i(t \cdot x) = h \cdot t_*(\xi_i(x))$, where h is a normalization function $\mathbb{R}^+ \cdot A_2 \to \mathbb{R}$ which ensures that the ξ_i lift $\frac{\partial}{\partial u_i}$ from U over the projection.

Finally, we define $\xi_i | \{0_P\} \times U = \frac{\partial}{\partial u_i}$.

Lemma 40. *The ξ_i are continuous and integrable, and stratified smooth with respect to $\{\{0\} \times U, F^{-1}\rho^{-1}(0) \setminus (\{0\} \times U), (N \times U) \setminus F^{-1}\rho^{-1}(0)\}$.*

Proof. Off $\mathbb{R}^+ \cdot (A_2 \cap A_1) \cup \{0_N\} \times U$ it is clear that the ξ_i are stratified smooth and continuous, since the original $\tilde{\xi}_i$ were stratified smooth and continuous, and we multiplied by a smooth function. Furthermore, the original $\tilde{\xi}_i$ are smooth off $F^{-1}\rho^{-1}(0)$, thus, in particular, their $\mathbb{R}^+$-extensions are smooth in a neighborhood of $\mathbb{R}^+ \cdot (A_2 \cap A_1)$. In a neighborhood of $\mathbb{R}^+ \cdot (A_2 \cap A_1)$ we know that, when restricting to the local product presentation, the vector field ξ_i is the unique one which has zero X- and $\mathbb{R}$-coordinate, and which must be smooth. Thus the ξ_i are smooth off $F^{-1}\rho^{-1}(0)$. The restrictions to $\{0_N\} \times U$ and $F^{-1}\rho^{-1}(0) \setminus (\{0_N\} \times U)$ are also smooth. Moreover, we see that the ξ_i are continuous off $\{0_N\} \times U$.

It remains to prove continuity at $\{0_N\} \times U$ and uniqueness of integral curves. Let (x_n, u_n) be a sequence in $W_N \subset N \times U$ converging to $(0, u) \in \{0_N\} \times U$, and consider the sequence

$$\xi_i(x_n, u_n) = (v_n, \frac{\partial}{\partial u_i}) \in T_{x_n}N \oplus T_{u_n}U = T_{x_n}N \oplus T_{u_n}\mathbb{R}^d.$$

We know that

$$\xi_i(x_n, u_n) = h(x_n, u_n) T t_n^{-1}(\bar{\xi}_i(t_n \cdot x_n, t_n \cdot u_n)),$$

where $\bar{\xi}_i$ is the vector field on A, $t_n \in \mathbb{R}^+$ such that $\rho(t_n \cdot (x_n, u_n)) = \epsilon$, and where h is a normalizing function to ensure that ξ_i lifts $\frac{\partial}{\partial u_i}$. Then $t_n^{-1} \to 0$.

Since $t_n \to \infty$ and the weights on $\mathbb{R}^d$ are non-positive, the set $\{\bar{\xi}_i(t_n \cdot (x_n, u_n)) | n \in \mathbb{N}\}$ is bounded. But then, if we denote

$$\bar{\xi}_i(t_n(x_n, u_n)) = \left((v_n^1, \ldots, v_n^m), \frac{\partial}{\partial u_i}\right) \in T_{t_n \cdot x_n}N \oplus T_{t_n \cdot u_n}\mathbb{R}^d$$

we see that

$$\begin{aligned} &h(x_n, u_n) T t_n^{-1}(\bar{\xi}_i(t_n \cdot (x_n, u_n))) \\ &= h(x_n, u_n)\left((t_n^{-1})^{a_1} v_1, \ldots, (t_n^{-1})^{a_m} v_m, (t_n^{-1})^{a_{m+i}} \cdot \frac{\partial}{\partial u_i}\right) \overset{n \to \infty}{\to} (0, \frac{\partial}{\partial u_i}), \end{aligned}$$

since, in particular, $h(x_n, u_n) \to 0$. But this was what we needed to prove for continuity of ξ_i.

Uniqueness of integral curves follows partly from the general lemma below, which ensures that the integral curves are unique off $\{0_N\} \times U$, and partly from the observation that all our integral curves off $\{0_N\} \times U$ are tangent to slices $t \cdot A$, and hence cannot reach the lower-dimensional stratum $\{0_N\} \times U$ in finite time.

Lemma 41. *Suppose that $g\colon X \to Y$ is a smooth, ST-stratified map and that we have a continuous vector field η on Y which is integrable and which is smooth on and tangent to strata. Let ξ be a continuous vector field on X, also smooth on and tangent to strata, which lifts η. Then ξ is integrable.*

Proof. Since ξ is smooth on and tangent to strata, it has a flow, whose restriction to any stratum is continuous and has unique integral curves. We show that the integral curves are globally unique by proving that they do not approach lower-dimensional strata in finite time.

Let $S_X \subset X$ be a stratum and let $S_Y \subset Y$ be the stratum such that $S_X = g^{-1}(S_Y)$. We show that there exists some neighborhood U of S_X in X such that no integral curve of ξ passing through a point of $U \setminus S_X$ will approach S_X in finite time. Since η is integrable, its integral curves are unique, and there exist a neighborhood V of S_Y which is invariant under the flow of η, and a continuous function $h\colon V \to \mathbb{R}$ which is constant on integral curves – it is easy to see that the existence of such a function is equivalent to the claim that the integral curves of η cannot approach S_Y in finite time.

But now $h \circ g$ is such a continuous function for $U = g^{-1}(V)$ and ξ, hence ξ has unique integral curves. Furthermore, ξ is continuous, hence it has a continuous flow by [CL55, II 4.4]. □

In our case the ST-invariant stratification is

$$\left(\{(\{0_P\} \times U), F^{-1}\rho^{-1}(0) \setminus (\{0_P\} \times U), (P \setminus 0) \times U\}, \{0 \times U, (P \setminus 0) \times U\}\right).$$

It follows that the ξ_i have unique integral curves. But continuous vector fields with unique integral curves are integrable [CL55, II, 4.4], and Lemma 40 holds. □

Now that we have constructed the continuous, integrable, stratified smooth vector fields $\{\eta_i\}$ and $\{\xi_i\}$, we construct the retractions R and S as usual. We see that S is E-tame, and that R is stratified smooth. □

Remark 42. The last statement, concerning stratified smooth vector fields in the source, could likely also be proven by extending the original argument by Looijenga [Loo77].

3. Multigerm equivalences

In this chapter we define what it means for a subgroup of $\mathscr{A}_f$ to be compact, where f is a multigerm. We show that when f is finitely $\mathscr{A}$-determined, there is a maximal such group, which is unique up to conjugacy. We denote an arbitrary representative of the conjugacy class by $MC(\mathscr{A}_f)$, and we show that the quotient $\mathscr{A}_f/MC(\mathscr{A}_f)$ is contractible, where contractibility is defined suitably.

All of this has been done for monogerms by K. Jänich [Jän78], C.T.C. Wall [Wal80] and R. Rimányi [Rim02], and in his thesis [Rim96, Theorem 1.6.3] Rimányi states the main theorem of our chapter in the case of stable multigerms without proof.

However, the multigerm case does not in any obvious way reduce to the monogerm case.

By Lemma 11, any stable multigerm $F\colon (\mathbb{R}^N, S) \to (\mathbb{R}^P, 0)$ with $|S| = s < \infty$ admits a decomposition

$$(43) \quad \left(\bigsqcup_{i=1}^{s} \sigma_i \circ (f_i \times \mathrm{id}_{\mathbb{R}^{P-p_i}})\right) \times \mathrm{id}_{\mathbb{R}^d} \colon \left(\bigsqcup_{i=1}^{s} \mathbb{R}^{n_i} \times \prod_{j \neq i} \mathbb{R}^{p_i}\right) \times \mathbb{R}^d \to \left(\prod_{i=1}^{s} \mathbb{R}^{p_i}\right) \times \mathbb{R}^d$$

in suitably chosen coordinates, where the f_i are ministable germs which are $\mathscr{E}\mathscr{K}$-equivalent to the germs of F at points in S. If F is ministable, then d will be 0, and in Rimányi's terminology, the f_i will be *roots of their kinds.*

It is natural to try to decompose the group $\mathscr{A}_F$ in terms of the groups $\mathscr{A}_{f_i}$, but there are some problems with such an approach. How do we know that there is only one way to choose the decomposition (43)? There might be several choices of coordinates giving such a decomposition, and corresponding to one choice of coordinates we might find elements of $\mathscr{A}_{f_1} \times \ldots \times \mathscr{A}_{f_s} \subset \mathscr{A}_f$ which do not belong to a product $\mathscr{A}_{f_1} \times \ldots \times \mathscr{A}_{f_s}$ after a change of coordinates. If the choice of coordinates *is* unique, then this is not trivial, and needs an explanation. If the choice of coordinates is *not* unique, then we cannot generally reduce to the monogerm case.

We had hoped that it might be easier to find a decomposition for compact subgroups $G < \mathscr{A}_f$, as these are conjugate to linear subgroups of $\mathscr{A}$. The subspaces $\mathbb{R}^{n_i}$ and $\mathbb{R}^{p_i}$ correspond to presentations of the multigerms $\prod_{j \neq i} f_i$, which must stay fixed under the action of $\mathscr{A}_f$, and linear actions are determined by the restricted action on subspaces spanning the whole source and target spaces. However, we meet yet another problem – can we find coordinates that *simultaneously* give f in the form (43) *and* linearize G? Suppose that we first put f in the form (43) and then linearize G; the linearization process might disturb the presentations by singularity type, which are linear subspaces of the target in (43), and take them into not-so-linear submanifolds. Another idea is to use the 1-jet of G, which is isomorphic to G – but unless f is linear, this might not leave f invariant, and hence $j^1 G$ might not belong to $\mathscr{A}_f$.

We shall, indeed, see that for statements concerning maximal compact subgroups of $\mathscr{A}_f$, we can reduce to the monogerm case – but this is not trivial. For statements concerning all of $\mathscr{A}_f$ – in particular concerning the contractability of the quotient $\mathscr{A}_f / MC(\mathscr{A}_f)$ – we need to reprove the theorems for multigerms.

Finally, we show that in some cases, which are of particular interest to us, the maximal compact subgroups $MC(\mathscr{A}_f)$ are very simple.

3.1. **Groups of multigerm equivalences and their maximal compact subgroups.**

Suppose given a multigerm

$$(44) \qquad f := f_1 \sqcup f_2 \sqcup \ldots \sqcup f_s \colon \underbrace{(\mathbb{R}^N, 0) \sqcup \ldots \sqcup (\mathbb{R}^N, 0)}_{s \text{ copies}} \to (\mathbb{R}^P, 0).$$

We have defined the groups $\mathscr{A}$ and $\mathscr{K}$ of multigerm equivalences in Chapter 1; let us repeat the definitions in Euclidean coordinates for the sake of notation.

Recall that $\mathscr{A}$ denotes the group of equivalences on the space of multigerms

$$C^\infty\left((\mathbb{R}^N,0)\sqcup\ldots\sqcup(\mathbb{R}^N,0);(\mathbb{R}^P,0)\right)$$

defined by $\bigsqcup_s \mathscr{R}\sqcup\mathscr{L}$, where $\mathscr{R}$ is the group of diffeomorphism germs of a source component $(\mathbb{R}^N,0)$, and $\mathscr{L}$ is the group of diffeomorphism germs of the target $(\mathbb{R}^P,0)$. We will write elements of $\mathscr{A}$ either in the form $\phi=(\phi_1,\ldots,\phi_s,\phi_t)$ where the ϕ_i are the diffeomorphisms in question, or sometimes in the form $\psi=(\psi_1,\ldots,\psi_s)$ where $\psi_i=\phi_i\times\phi_t$ for each i, when we consider ψ as an element in $\mathscr{K}$ as defined below.

The group $\mathscr{K}$ of equivalences on the same space of multigerms is the group of diffeomorphism germs

$$H\colon\left(\bigsqcup_s\mathbb{R}^N\times\mathbb{R}^P,\bigsqcup_s(0,0)\right)\to\left(\bigsqcup_s\mathbb{R}^N\times\mathbb{R}^P,\bigsqcup_s(0,0)\right)$$

such that the diagram below commutes:

$$\begin{array}{ccccc}
\bigsqcup_s(\mathbb{R}^N,0) & \xrightarrow{\mathrm{id}\times 0} & \bigsqcup_s\left(\mathbb{R}^N\times\mathbb{R}^P,(0,0)\right) & \xrightarrow{\mathrm{pr}_{\mathbb{R}^N}} & \bigsqcup_s(\mathbb{R}^N,0)\\
\Big\downarrow{\scriptstyle H_0} & & \Big\downarrow{\scriptstyle H} & & \Big\downarrow{\scriptstyle H_0}\\
\bigsqcup_s(\mathbb{R}^N,0) & \xrightarrow[\mathrm{id}\times 0]{} & \bigsqcup_s\left(\mathbb{R}^N\times\mathbb{R}^P,(0,0)\right) & \xrightarrow[\mathrm{pr}_{\mathbb{R}^N}]{} & \bigsqcup_s(\mathbb{R}^N,0)
\end{array}$$

where $H_0=H|\mathbb{R}^N\times\{0\}$. We will write elements of $\mathscr{K}$ in the form $\psi=(\psi_1,\ldots,\psi_s)$ where, in fact, $\psi=\bigsqcup_i\psi_i$.

Recall that $\mathscr{K}$ acts on germs f in the following way: $H\cdot f=g$ if $(\mathrm{id},f)\circ H_0=H\circ(\mathrm{id},g)$.

Given a multigerm f, we denote by $\mathscr{A}_f$ and $\mathscr{K}_f$ the *isotropy subgroups* of $\mathscr{A}$ and $\mathscr{K}$ at f; namely the subgroup of elements in $\mathscr{A}$ and $\mathscr{K}$, respectively, that leave f invariant when viewed as groups acting on the space of multigerms.

Since the action of the $\mathscr{K}$-group on source and target is actually an action on a disjoint union of spaces, one for each component f_i of the germ f, we can decompose the groups $\mathscr{K}$ and $\mathscr{K}_f$ for multigerms to a product of $\mathscr{K}$-groups for monogerms:

Proposition 45. *The group* $\mathscr{K}$ *acting on*

$$C^\infty\left((\mathbb{R}^N,0)\sqcup\ldots\sqcup(\mathbb{R}^N,0);(\mathbb{R}^P,0)\right)$$

is given by

$$\underbrace{\tilde{\mathscr{K}}\times\ldots\times\tilde{\mathscr{K}}}_{s\text{ copies}},$$

where $\tilde{\mathscr{K}}$ *is the group acting on* $C^\infty\left((\mathbb{R}^N,0),(\mathbb{R}^P,0)\right)$, *and the isotropy subgroup*

$$\mathscr{K}_f$$

is given by

$$\tilde{\mathscr{K}}_{f_1}\times\ldots\times\tilde{\mathscr{K}}_{f_s}.$$

Proof. This follows directly from the definition. □

The analogous argument does not work for groups of $\mathscr{A}$-equivalences, however, as we couple a family of diffeomorphism germs in the source with only *one* diffeomorphism germ in the target, which causes problems with the isotropy subgroups. Consequently, the main objective of this section is to prove that $\mathscr{A}_f$ really does have a maximal compact subgroup when f is a finitely $\mathscr{A}$-determined multigerm.

3.1.1. *Compact subgroups: definition and properties.* We define a compact subgroup of $\mathscr{A}$ to be a subgroup G of $\mathscr{A}$ which is conjugate in $\mathscr{A}$ to a compact subgroup of $\bigsqcup_s GL_N\sqcup GL_P<\mathscr{A}$. The definition is reasonable: Suppose that $G<\mathscr{A}$ is isomorphic to some compact Lie group $\tilde{G}$, such that $\tilde{G}$ acts diffeomorphically on $\bigsqcup_s\mathbb{R}^N\sqcup\mathbb{R}^P$ through the isomorphism $\tilde{G}\to G$, keeping the origins fixed (compare with the definition of [dPW]). Then, in particular, $\tilde{G}$ acts diffeomorphically on each of the $\mathbb{R}^N$ and on $\mathbb{R}^P$. By Bochner's theorem we can choose local coordinates in $(\mathbb{R}^N,0),\ldots,(\mathbb{R}^N,0)$ and $(\mathbb{R}^P,0)$, respectively, given by diffeomorphism germs $\phi_1,\ldots,\phi_s$ and ϕ_t, with respect to which G acts linearly on the $\mathbb{R}^N$ and on $\mathbb{R}^P$. These define an element $(\phi_1,\ldots,\phi_s,\phi_t)$ of $\mathscr{A}$, which linearizes G in $\mathscr{A}$.

By analogy, we define a compact subgroup of $\mathscr{K}$ to be a subgroup $G<\mathscr{K}$ which is conjugate in $\mathscr{K}$ to a compact linear subgroup of $\bigsqcup_s GL_N\times GL_P$.

We start out by making a couple of trivial observations concerning the group of $\mathscr{A}$-equivalences as a subgroup of the $\mathscr{K}$-equivalences:

Lemma 46.

i) *We have* $\mathscr{A}_f<\mathscr{K}_f$, *so in particular* $MC(\mathscr{A}_f)<MC(\mathscr{K}_f)$, *whenever the maximal compact subgroups exist.*
ii) *Suppose that* $f\colon(\mathbb{R}^N,0)\to(\mathbb{R}^P,0)$ *is a monogerm, and consider* $\mathscr{A}$ *and* $\mathscr{K}$ *as groups acting on the source and target of* f. *Then* $\mathscr{K}\cap GL_{N+P}$ *is a subgroup of* $\mathscr{A}$, *so in particular,* $\mathscr{K}_f\cap GL_{N+P}<\mathscr{A}_f$. □

Remark 47. Note that Lemma 46 does *not* generally imply that $MC(\mathscr{K}_f)=MC(\mathscr{A}_f)$ for a monogerm f. Suppose that G is a maximal compact subgroup of $\mathscr{K}_f$; then there exist a group $\tilde{G}=gGg^{-1}$ conjugate to G, and a map $\tilde{f}=g\cdot f$ which is $\mathscr{K}$-equivalent to f, such that $\tilde{G}<\mathscr{K}_{\tilde{f}}$ and $\tilde{G}<GL_{N+P}$. Then $\tilde{G}<\mathscr{A}_{\tilde{f}}$, but (!) this does not mean that $G<\mathscr{A}_f$, since the conjugating group element $g\in\mathscr{K}$ does not necessarily belong to $\mathscr{A}$.

Throughout the rest of this section, we let $\mathscr{H}$ denote $\mathscr{A}$ or $\mathscr{K}$ unless otherwise is specified.

Lemma 48. *Suppose that a multigerm $f\colon \bigsqcup_s(\mathbb{R}^N,0)\to(\mathbb{R}^P,0)$ is finitely $k-\mathscr{H}$-determined, and suppose that $G<\mathscr{H}_f$ is compact. Then we can find an element $\phi\in\mathscr{H}$ and a subgroup $\tilde{G}$ of $\mathscr{H}_{\phi\cdot f}$ such that $\phi\cdot f$ is a polynomial of degree $\leq k$, and $\tilde{G}$ is a linear group which is conjugate (in $\mathscr{H}$) to G.*

Proof. Since $G<\mathscr{H}_f$ is compact, we can find $\psi\in\mathscr{H}$ such that $\psi G\psi^{-1}$ is linear, by definition. Now $\tilde{G}=\psi G\psi^{-1}<\mathscr{H}_{\psi\cdot f}$, and since f is $k-\mathscr{H}$-determined, so is $\psi\cdot f$. In particular, $\psi\cdot f$ is $\mathscr{H}$-equivalent to the polynomial representative p of its k-jet $j^k(\psi\cdot f)$, by an element $\tilde{\psi}\in\mathscr{H}$, say:

$$p=\tilde{\psi}\cdot\psi\cdot f.$$

We claim that $\tilde{G}<\mathscr{H}_p$. Since the linear group $\tilde{G}$ leaves $\psi\cdot f$ invariant, it certainly leaves the jet $j^k(\psi\cdot f)$ invariant in $J^k(n,p)$. But then it must leave p invariant, since then, for any $g\in\tilde{G}$, the map $g\cdot p$ is a degree k polynomial representing $j^k(\psi\cdot f)$.

Set $\phi=\tilde{\psi}\cdot\psi$, and we are done. □

Lemma 49. *For any compact subgroup G of $\mathscr{H}$ the restriction of the 1-jet map $j^1|\colon G\to\bigsqcup_s GL_N\times GL_P$ is injective.*

Proof. By the definition of a compact subgroup there exists a choice of coordinates on $\bigsqcup_s\mathbb{R}^N\times\mathbb{R}^P$ such that G acts linearly; now in these coordinates the 1-jet map is just the inclusion into $\bigsqcup_s GL_N\times GL_P$. The topological properties of the map j^1 do not depend on the choice of coordinates; hence $j^1|G$ is injective. □

3.1.2. *Maximal compact subgroups: Existence and uniqueness.* Now we are ready to state and prove the main theorem of the section.

Theorem 50. *Let f be a finitely $\mathscr{H}$-determined multigerm as in (44). The group $\mathscr{H}_f$ has a maximal compact subgroup, which is unique up to conjugation in $\mathscr{H}_f$.*

Remark 51. The monogerm version of this theorem was proven by Jänich [Jän78] (for $\mathscr{H}=\mathscr{R}$) and Wall [Wal80] (for $\mathscr{H}=\mathscr{A}$ or $\mathscr{K}$) with some completing comments by du Plessis and Wilson [dPW, p. 270], who proved similar results for actions of $\mathscr{R}$, but without finite $\mathscr{R}$-determinacy.

Proof. Just like in the monogerm case we will need an equivariant finite determinacy condition:

Lemma 52. *Let $f\colon\bigsqcup_s(\mathbb{R}^N,0)\to(\mathbb{R}^P,0)$ be $k-\mathscr{H}$-determined, and suppose that G is a compact linear subgroup of $\mathscr{H}_f$. Then f is $k-G$-determined; that is, if $\tilde{f}\colon\bigsqcup_s(\mathbb{R}^N,0)\to(\mathbb{R}^P,0)$ is G-invariant such that $j^kf=j^k\tilde{f}$, then there exists $\phi\in\mathscr{H}$ such that $\phi\cdot g=g\cdot\phi$ for all $g\in G$, and $\tilde{f}=\phi\cdot f$. In particular, $G<\mathscr{H}_{\tilde{f}}$, since $g\cdot\tilde{f}=g\cdot\phi\cdot f=\phi\cdot g\cdot f=\phi\cdot f=\tilde{f}$ for all $g\in G$.*

Proof. The proof follows the standard proof of k-determinacy [Mat68a], using vector fields – but averaging the vector fields over the group G using the Haar integral. □

Assume that f is $k-\mathscr{H}$-determined. We write $\mathscr{H}^k$ for the families of invertible k-jets

$$\phi^k = (\phi_1^k, \dots, \phi_s^k) \colon \left(\bigsqcup_s \mathbb{R}^N \times \mathbb{R}^P, \bigsqcup_s 0\right) \to \left(\bigsqcup_s \mathbb{R}^N \times \mathbb{R}^P, \bigsqcup_s 0\right), \quad \phi \in \mathscr{H},$$

and write $\mathscr{H}_f^k$ for the subgroup stabilizing $j^k f$; both these groups are real algebraic.

Real algebraic groups have finitely many components, so we may apply Iwasawa's theorem [Hoc65, p. 180] and choose a maximal compact subgroup G of $\mathscr{H}_f^k$. Following the arguments of Jänich [Jän78, §2] and Bochner [Boc45] we linearize G in the following way:

Lemma 53. *Suppose that $G < \mathscr{H}^k$ is a compact subgroup. Then there exists $\phi \in \mathscr{H}^k$ such that $\phi G \phi^{-1} < \bigsqcup_s GL_N \times GL_P$.*

Proof. Using the Haar integral over G we average the family of identity jets in $\mathscr{H}^k$ in order to make them equivariant with respect to $G < \mathscr{H}^k$ and $j^1 \colon G \to \mathscr{H}^k$. More precisely, we set

$$\tilde{\phi} := \int_G j^1(g)\rho(g)^{-1} dg \in \ker(j^1 \colon \mathscr{H} \to GL_{sN+P}),$$

where ρ is the homomorphism from G to $\mathscr{H}$ taking the elements of $\mathscr{H}^k$ to their polynomial representatives in $\mathscr{H}$, and define $\phi = j^k\tilde{\phi}$. Then $j^1 g = \phi g \phi^{-1}$ for all $g \in G$. □

Still following Jänich, we can prove a strong linearization theorem for compact subgroups of $\mathscr{H}$:

Lemma 54. *Suppose that $G < \mathscr{H}$ is a compact subgroup. Then there exists $\phi \in \mathscr{H}$ such that $\phi G \phi^{-1} < \bigsqcup_s GL_N \times GL_P$. Suppose, furthermore, that $j^k G < \bigsqcup_s GL_N \times GL_P < \mathscr{H}^k$. Then we may assume that $j^k \phi = (1, \dots, 1)$.*

Proof. As in [Jän78, §2], we define $\phi = \int_G (j^1 g) g^{-1} dg$, and note that if $j^k g$ is linear for each $g \in G$, then $j^k \phi$ is identity. □

We return to the proof of Theorem 50, and to the maximal compact subgroup G of $\mathscr{H}_f^k$.

Denote $G_0 = \phi G \phi^{-1} < \bigsqcup_s GL_N \times GL_P$. By abuse of notation, we will also denote by G_0 the corresponding linear subgroup of $\mathscr{H}^k$. Let $\tilde{\phi} \in \mathscr{H}$ such that $j^k\tilde{\phi} = \phi$. Then if $f_0 = \tilde{\phi} \cdot f$, its jet $j^k f_0$ is G_0-invariant, and G_0 is maximal compact in $\mathscr{H}_{f_0}^k$.

Let H be any compact subgroup of $\mathscr{H}_{f_0}$. By classical Lie group theory $j^k H$ is conjugate in $\mathscr{H}_{f_0}^k$ to a subgroup of G_0, say by a family of jets of diffeomorphisms $\psi^k = (\psi_1^k, \dots, \psi_s^k)$. Let $\tilde{\psi}$ be a family of diffeomorphisms $(\tilde{\psi}_1, \dots, \tilde{\psi}_s)$ with jet ψ^k. Then the k-jet of $\tilde{\psi} H \tilde{\psi}^{-1}$ is linear.

By Lemma 54, we can find $\psi = (\psi_1, \dots, \psi_s)$ with the same k-jet as $\tilde{\psi}$, such that $\psi H \psi^{-1}$ is linear. Hence $\tilde{f} = \psi \cdot f_0$ is $(\psi H \psi^{-1})$-invariant, just like f_0 (because

$\psi H\psi^{-1} < G_0$). Since $j^k\psi = j^k\tilde{\psi} = \psi^k \in \mathscr{H}^k_{f_0}$, we have $j^k\tilde{f} = j^k f_0$. Recall that f_0 is $k-(\psi H\psi^{-1})$-determined, so by Lemma 52 there exists some $(\psi H\psi^{-1})$-equivariant family of diffeomorphisms $\alpha = (\alpha_1, \ldots, \alpha_s) \in \mathscr{H}$ such that $f_0 = \alpha \cdot \tilde{f}$. Hence $f_0 = (\alpha\psi) \cdot f_0$, and so $\alpha\psi$ preserves f_0 *and* conjugates H to

$$\alpha \underbrace{\psi H \psi^{-1}}_{<G_0} \alpha^{-1} = \psi H \psi^{-1} < G_0$$

where the last equality holds because α is G_0-equivariant.

We have seen that G_0 can be viewed as a compact subgroup of $\mathscr{H}_{f_0}$, and that any other compact subgroup H of $\mathscr{H}_{f_0}$ is conjugate to a subgroup of G_0, so conjugating back to $\mathscr{H}_f$, we see that Theorem 50 holds. □

3.2. **The quotient $\mathscr{A}_f/MC(\mathscr{A}_f)$ is contractible.** Following Jänich [Jän78], we define what it means for $\mathscr{A}_f/MC(\mathscr{A}_f)$ to be contractible. Note that we have not specified a topology on $\mathscr{A}_f/MC(\mathscr{A}_f)$, and in fact we shall define contractibility not in terms of the topology of $\mathscr{A}_f/MC(\mathscr{A}_f)$ as a space of its own, but through the topological properties of its action on the source and target of f.

Before defining contractibility, we need to decide what it means for a map into a quotient $\mathscr{A}_f/G$ to be smooth. Let G be a subgroup of $\mathscr{A}$. Given a smooth manifold M, possibly with boundary, we say that a map $q\colon M \to \mathscr{A}/G$ is *smooth* if there exists an open covering $\{U_i\}$ of M such that q is represented by fibered (over U_i) maps $\phi_i\colon U_i \times \bigsqcup_s \mathbb{R}^n \to U_i \times \bigsqcup_s \mathbb{R}^n$ and $\psi_i\colon U_i \times \mathbb{R}^p \to U_i \times \mathbb{R}^p$ which are diffeomorphisms.

Equivalently, a map $\alpha\colon M \to \mathscr{A}/G$ is smooth if there exists an open covering $\{U_i\}_{i\in I}$ of M such that α admits a local lift $\tilde{\alpha}_i\colon U_i \to \mathscr{A}$, such that the corresponding fibered map-germs ϕ_i and ψ_i are smooth.

Definition 55. Let G be a subgroup of $\mathscr{A}_f$. The quotient $\mathscr{A}_f/G$ is *contractible* if for every smooth manifold M with boundary, any smooth map $\partial M \to \mathscr{A}_f/G$ can be extended to a smooth map $M \to \mathscr{A}_f/G$.

We proceed to state and prove the main result of the section:

Theorem 56. *Suppose given a finitely $\mathscr{A}$-determined multigerm*

$$f = \bigsqcup_{i=1}^{s} f_i\colon \bigsqcup_{i=1}^{s} (\mathbb{R}^N, 0) \to (\mathbb{R}^P, 0).$$

Then the quotient $\mathscr{A}_f/MC(\mathscr{A}_f)$ is contractible.

This proof follows that of Rimányi [Rim96] for the monogerm case.

Proof. The following proposition is crucial to the proof:

Proposition 57. *There exists an $l \in \mathbb{N}$ such that the following holds:*

If M is an r-dimensional manifold with boundary (possibly empty) and

$$g, h\colon (M \times \bigsqcup_s \mathbb{R}^N, M \times \bigsqcup_s \{0\}) \to (M \times \mathbb{R}^P, M \times \{0\})$$

are fibered (over M) germs at $M \times \bigsqcup_s \{0\}$ satisfying the following properties:

$$g|\partial M \times \bigsqcup_s (\mathbb{R}^N, 0) = h|\partial M \times \bigsqcup_s (\mathbb{R}^N, 0), \text{ and}$$

$$j^l(g|u \times \mathbb{R}^N) = j^l(h|u \times \mathbb{R}^N) = j^l f \quad \forall u \in M,$$

then there exist $\psi^k \in \mathrm{Diff}(M \times \mathbb{R}^N)$, $k = 1, \dots, s$, and $\phi \in \mathrm{Diff}(M \times \mathbb{R}^P)$ such that $g = \phi \circ h \circ \bigsqcup_s (\psi^k)^{-1}$ and

$$\psi^k|\partial M \times \mathbb{R}^N = \mathrm{id}, \qquad \phi|\partial M \times \mathbb{R}^P = \mathrm{id},$$

$$j^1(\psi^k|u \times \mathbb{R}^N) = \mathrm{id}, \qquad j^1(\phi|u \times \mathbb{R}^P) = \mathrm{id},$$

for all $u \in M$, $k = 1, \dots, s$.

Remark 58. Paraphrased, Proposition 57 says that there exists a smooth map $\varphi \colon M \to \mathscr{A}$ such that $\varphi|\partial M \equiv \mathrm{id}$, $j^1\varphi \equiv \mathrm{id}$ and $g = \varphi \cdot h$.

Proof. We find the ψ^k and ϕ by using the flows of suitably chosen vector fields in the source components and in target.

Let

$$F\colon \left(M \times \bigsqcup_s \mathbb{R}^N \times \mathbb{R}, M \times \bigsqcup_s \{0\} \times \mathbb{R} \right) \to (M \times \mathbb{R}^P \times \mathbb{R}, M \times \{0\} \times \mathbb{R})$$

be the map germ defined by

$$(u, x, t) \mapsto ((1-t)g(u, x) + th(u, x), t)\,.$$

From now on we denote by

$$u = (u_i), \quad x^k = (x_i^k), \quad y = (y_i), \quad t,$$

the coordinates of M, the kth source component $\mathbb{R}^N$, $\mathbb{R}^P$, and $\mathbb{R}$. We write $F = \bigsqcup_{k=1}^s F^k$, and the notation F_y will denote the composition $\mathrm{pr}_{\mathbb{R}^P} \circ F$ and so on.

3.2.1. *Constructing the diffeomorphisms.* We want to construct flows Ψ and Φ in source and target such that

$$\begin{aligned} &\Psi| \colon (M \times \textstyle\bigsqcup_s \mathbb{R}^N \times \mathbb{R}) \times [0,1] \to M \times \textstyle\bigsqcup_s \mathbb{R}^N \times \mathbb{R}, \quad \Psi = \textstyle\bigsqcup_{k=1}^s \Psi^k \\ &\Phi| \colon (M \times \mathbb{R}^P \times \mathbb{R}) \times [0,1] \to M \times \mathbb{R}^P \times \mathbb{R} \end{aligned}$$

with

$$\begin{aligned} &\Psi^k\left((u, x^k, 0), s\right) \in M \times \mathbb{R}^N \times \{s\}, \\ &\Phi\left((u, y, 0), s\right) \in M \times \mathbb{R}^P \times \{s\}, \end{aligned}$$

for all $s \in [0,1]$, and

$$F(\Psi((u, x, 0), s)) = \Phi((g(u, x), 0), s) = \Phi(F(u, x, 0), s),$$

which holds if

$$F \circ \Psi = \Phi \circ (F \times \mathrm{pr}_{[0,1]}). \tag{59}$$

Suppose that we have found such flows Ψ and Φ, and define maps

$$\begin{aligned}\tilde{\Psi}&\colon M\times\bigsqcup_s\mathbb{R}^N\times[0,1]\to M\times\bigsqcup_s\mathbb{R}^N\\ \tilde{\Phi}&\colon M\times\mathbb{R}^P\times[0,1]\to M\times\mathbb{R}^P\end{aligned}$$

by setting

$$\begin{aligned}\tilde{\Psi}(u,x,s)&=\mathrm{pr}_{M\times\bigsqcup_s\mathbb{R}^N}\circ\Psi((u,x,0),s)\\ \tilde{\Phi}(u,y,s)&=\mathrm{pr}_{M\times\mathbb{R}^P}\circ\Phi((u,y,0),s),\end{aligned}$$

and define

$$\tilde{h}\colon M\times\bigsqcup_s\mathbb{R}^N\times[0,1]\to M\times\mathbb{R}^P$$

by setting

$$\tilde{h}(u,x,s)=\left(\tilde{\Phi}_s^{-1}\circ h\circ\tilde{\Psi}\right)(u,x,s),$$

where $\tilde{\Phi}_s(u,y)=\tilde{\Phi}(u,y,s)$. Note, in particular, that $\tilde{\Phi}_0(u,y)=(u,y)$, and that $\tilde{\Psi}(u,x,0)=(u,x)$.

Lemma 60. *Then $\tilde{h}_0=h$ and $\tilde{h}_1=g$.*

Proof. It is easy to see that $\tilde{h}_0=h$:

$$\begin{aligned}\tilde{h}_0(u,x)&=\tilde{h}(u,x,0)\\ &=(\tilde{\Phi}_0^{-1}\circ h\circ\tilde{\Psi})(u,x,0)\\ &=(\tilde{\Phi}_0^{-1}\circ h)(u,x)\\ &=h(u,x).\end{aligned}$$

For the second identity we note that

$$\tilde{h}_1(u,x)=\tilde{h}(u,x,1)=\tilde{\Phi}_1^{-1}(h(\tilde{\Psi}(u,x,1))),$$

so $\tilde{h}_1=g$ if and only if $(\tilde{\Phi}_1\circ g)(u,x)=(h\circ\tilde{\Psi})(u,x,1)$ for all u,x, which holds if and only if

$$\tilde{\Phi}(g(u,x),1)=h(\tilde{\Psi}(u,x,1))\text{ for all }u,x,\tag{61}$$

where $h(\tilde{\Psi}(u,x,1))=h(\mathrm{pr}_{M\times\bigsqcup_s\mathbb{R}^n}(\Psi((u,x,0),1)))$. But

$$F\left(\Psi\left((u,x,0),1\right)\right)=\Phi\left(F(u,x,0),1\right)$$

by (59), and $\Phi\left(F(u,x,0),1\right)=\Phi(g(u,x),0,1)$ by the definition of F, while

$$F\left(\Psi\left((u,x,0),1\right)\right)=\left(h\left(\mathrm{pr}_{M\times\bigsqcup_s\mathbb{R}^n}(\Psi((u,x,0),1))\right),1\right),$$

also by the definition of F, so

$$\begin{aligned}h(\tilde{\Psi}(u,x,1))&=\mathrm{pr}_{M\times\mathbb{R}^p}(F(\Psi((u,x,0),1)))\\ &=\mathrm{pr}_{M\times\mathbb{R}^p}(\Phi(g(u,x),0,1))\\ &=\tilde{\Phi}(g(u,x),1),\end{aligned}$$

and (61) holds. ∎

In particular, $\tilde{\Phi}_1^{-1} \circ h \circ \tilde{\Psi}_1 = \tilde{h}_1 = g$, and thus $\tilde{\Psi}_1$ and $\tilde{\Phi}_1$ are the wanted conjugating diffeomorphisms $\bigsqcup \psi^k$ and ϕ.

Suppose that we are given map germs

$$X^k\colon (M \times \mathbb{R}^N \times \mathbb{R}, M \times 0 \times \mathbb{R}) \to (\mathbb{R}^N, 0) \; (k = 1, \dots, s)$$
$$Y\colon (M \times \mathbb{R}^P \times \mathbb{R}, M \times 0 \times \mathbb{R}) \to (\mathbb{R}^P, 0)$$

such that the following conditions (62) – (65) hold:

$$\sum_{i=1}^{n} \frac{\partial F_{y_j}}{\partial x_i}(u, x^k, t) X_i^k(u, x^k, t) + \frac{\partial F_{y_j}}{\partial t}(u, x^k, t) = Y_j(F(u, x^k, t)) \tag{62}$$

for all $j = 1, \dots, P$, and $k = 1, \dots, s$;

$$X^k | M \times 0 \times \mathbb{R} = 0 \; (k = 1, \dots, s); \qquad Y | M \times 0 \times \mathbb{R} = 0; \tag{63}$$

$$\begin{cases} \frac{\partial X^k}{\partial x_i}(u, 0, t) = 0 \text{ for all } i = 1, \dots, N, \quad (k = 1, \dots, s); \\ \frac{\partial Y}{\partial y_j}(u, 0, t) = 0 \text{ for all } j = 1, \dots, P; \end{cases} \tag{64}$$

$$X^k | \partial M \times \mathbb{R}^N \times \mathbb{R} = 0; \qquad Y | \partial M \times \mathbb{R}^P \times \mathbb{R} = 0. \tag{65}$$

Consider the flows of the following vector fields:

$$\tilde{X}^k\colon M \times \mathbb{R}^N \times \mathbb{R} \to TM \times T\mathbb{R}^N \times T\mathbb{R}, \quad (u, x^k, t) \mapsto (0, X(u, x^k, t), 1),$$
$$\tilde{Y}\colon M \times \mathbb{R}^P \times \mathbb{R} \mapsto TM \times T\mathbb{R}^P \times T\mathbb{R}, \quad (u, y, t) \mapsto (0, Y(u, y, t), 1).$$

By (63) these flows exist at least in a neighborhood of $M \times 0 \times \mathbb{R}$. Furthermore, we see that the condition (62) is just the condition of being a derivative of F-related flows, namely satisfying (59).

The maps $\tilde{\Psi}_1 = \bigsqcup_{k=1}^{s} \tilde{\Psi}_1^k$ and $\tilde{\Phi}_1$ associated with the flows of $\tilde{X}^k$ and $\tilde{Y}$ as described above clearly satisfy

$$\tilde{\Psi}_1^k | \partial M \times \mathbb{R}^N = \mathrm{id} \quad \tilde{\Phi}_1 | \partial M \times \mathbb{R}^P = \mathrm{id}$$
$$j^1(\tilde{\Psi}_1^k | u \times \mathbb{R}^N) = \mathrm{id} \quad j^1(\tilde{\Phi} | u \times \mathbb{R}^P) = \mathrm{id}$$

by (64) and (65). Hence, if we can find such vector fields X^k and Y, we are done.

3.2.2. *It suffices to find* local *(in M) vector fields X^k and Y satisfying (62) – (65).* It is enough to find the vector fields locally because the conditions are convex on fibers, and we can piece local solutions together using a partition of unity.

We divide the problem into two parts:

Case I Find the local solution near $(u, t) \in \mathrm{int} M \times [0, 1]$, and
Case II Find the local solution near $(u, t) \in \partial M \times [0, 1]$.

3.2.3. *Case I.* Here (65) is meaningless, and (62) – (64) can be summarized in the condition that

$$\left(\left(\frac{\partial F_{y_j}}{\partial t}(u, x^k, t)\right)_{j=1}^{P}\right)_{k=1}^{s}$$

belongs to

$$\bigoplus_{k=1}^{s} m(N)^2 \left\langle \left(\frac{\partial F_{y_1}^k}{\partial x_i}, \dots, \frac{\partial F_{y_P}^k}{\partial x_i}\right) \middle| \, 1 \leq i \leq N \right\rangle_{\mathscr{E}(r+N+1)} + F^*(m(P)^2\mathscr{E}(r+P+1))^P.$$

Note that we identify the module of map-germs $\bigsqcup_{k=1}^{s}(\mathbb{R}^n, 0) \to \mathbb{R}^p$ with the direct sum $\bigoplus_{k=1}^{s} \mathscr{E}(n,p) = \bigoplus_{k=1}^{s} \left(\bigoplus_p \mathscr{E}(p)\right)$, and so on.

Any coordinate of the left is of the form

$$\left(\frac{\partial F_{y_j}}{\partial t}(u, x^k, t)\right)_{k=1}^{s} \sim \bigsqcup_{k=1}^{s} \frac{\partial F_{y_j}}{\partial t}(u, x^k, t) = h_{y_j}(u,x) - g_{y_j}(u,x)$$

and lies in $\bigoplus_{k=1}^{s} m(N)^{l+1}\mathscr{E}(r+N+1)^N$, since g and h have the same l-jets in each fiber.

Thus it suffices to show

$$\begin{aligned} &\bigoplus\nolimits_{k=1}^{s} m(N)^{l+1}\mathscr{E}(r+N+1)^P \\ &\subset tF(\bigoplus\nolimits_{k=1}^{s} m(N)^2\mathscr{E}(r+N+1)^N) + wF(m(P)^2\mathscr{E}(r+P+1)^P), \end{aligned} \tag{66}$$

where tF and wF are defined as on p. 12.

Since f is finitely $\mathscr{A}$-determined, we know that

$$\bigoplus_{k=1}^{s} m(N)^q\mathscr{E}(N)^P \subset tf(\bigoplus_{k=1}^{s} m(N)^2\mathscr{E}(N)^N) + wf(m(P)^2\mathscr{E}(P)^P) \tag{67}$$

for some $q \in \mathbb{N}$ [Mat68a, Chapter 6, Lemma 2].

3.2.4. *(67) implies (66).* First, we compare the right hand sides of (66) and (67). The following diagram, where the vertical maps are induced by inclusions, does not commute:

$$\begin{array}{ccccc}
\bigoplus_{k=1}^{s} \mathscr{E}(N)^N & \xrightarrow{tf} & \mathscr{E}(N)^P & \xleftarrow{wf} & \mathscr{E}(P)^P \\
\downarrow & & \downarrow & & \downarrow \\
\bigoplus_{k=1}^{s} \mathscr{E}(r+N+1)^N & \xrightarrow[tF]{} & \mathscr{E}(r+N+1)^P & \xleftarrow[wF]{} & \mathscr{E}(r+P+1)^P
\end{array}$$

but we can say something about how far from commuting it really is.

Lemma 68. *If $\varphi \in \mathscr{E}(P)^P$, then $wF(\varphi) - wf(\varphi) \in \bigoplus_{k=1}^{s} m(N)^{l+1}\mathscr{E}(r+N+1)^P$.*

Proof. Since $wF(\varphi) = \bigsqcup_{k=1}^{s} \varphi \circ F^k$ and similarly for f, it suffices to show that the corresponding monogerm claim holds separately for each F^k. This goes as in [Rim02, Lemma 2.3]. □

Lemma 69. *If $\varphi = (\varphi_k)_{k=1}^{s} \in \bigoplus_{k=1}^{s} \mathscr{E}(N)^N$ then $tF(\varphi) - tf(\varphi) \in \bigoplus_{k=1}^{s} m(N)^l \cdot \mathscr{E}(r+N+1)^P$.*

Proof. Since $tF((\varphi_k)_{k=1}^{s}) = \bigsqcup_{k=1}^{s} TF^k \circ \varphi_k = \bigsqcup_{k=1}^{s} tF^k(\varphi_k)$ and similarly for f, it suffices to show that the claim holds separately for each F^k. This goes as in [Rim02, Lemma 2.4]. □

Denote by U the intersection

$$wf^{-1}\left(tf\left(\bigoplus_{k=1}^{s} m(N)^2\mathscr{E}(N)^N\right) + \bigoplus_{k=1}^{s} m(N)^q\mathscr{E}(N)^P\right) \cap m(P)^2\mathscr{E}(P)^P,$$

where q is the number from (67). This is an $\mathscr{E}(P)$-submodule of $\mathscr{E}(P)^P$. Denote by V the $\mathscr{E}(r+P+1)$-submodule of $\mathscr{E}(r+P+1)^P$ generated by the image of U under the natural inclusion $\mathscr{E}(P)^P \hookrightarrow \mathscr{E}(r+P+1)^P$.

3.2.5. *We claim that:*

$$\begin{aligned} &wF(V) + tF(\textstyle\bigoplus_{k=1}^{s} m(N)^2\mathscr{E}(r+N+1)^N) \\ (70)\qquad &+ \textstyle\bigoplus_{k=1}^{s} m(N)^q m(r+N+1)^{l-q}\mathscr{E}(r+N+1)^P \\ &= tF(\textstyle\bigoplus_{k=1}^{s} m(N)^2\mathscr{E}(r+N+1)^N) + \bigoplus_{k=1}^{s} m(N)^q\mathscr{E}(r+N+1)^P. \end{aligned}$$

Note that in order to show "$\subset$", it is enough to show

$$wF(U) \subset RHS,$$

since RHS is an $\mathscr{E}(r+P+1)$-submodule of θ_F.

Let $v \in U$; then $v = wf^{-1}(tf(\xi) + \zeta)$ for some $\xi \in \bigoplus_{k=1}^{s} m(N)^2\mathscr{E}(N)^N$, $\zeta \in \bigoplus_{k=1}^{s} m(N)^q\mathscr{E}(N)^P$.

Then

$$wF(v) = \overbrace{(wF(v) - wf(v))}^{(*)} + \overbrace{(tf(\xi) - tF(\xi))}^{(**)} + \zeta + tF(\xi).$$

By Lemmas 68 and 69, the element $(*)$ lies in $\bigoplus_{k=1}^{s} m(N)^{l+1}\mathscr{E}(r+N+1)^P$, and the element $(**)$ lies in $\bigoplus_{k=1}^{s} m(N)^l\mathscr{E}(r+N+1)^P$. Choosing $l > q$ in the statement of the proposition, we have

$$wF(v) \in \bigoplus_{k=1}^{s} m(N)^q\mathscr{E}(r+N+1)^P + tF(\bigoplus_{k=1}^{s} m(N)^2\mathscr{E}(r+N+1)^N),$$

giving $LHS \subset RHS$ in (70).

In order to prove $RHS \subset LHS$ in (70), it is enough to prove

$$\bigoplus_{k=1}^{s} m(N)^q\mathscr{E}(r+N+1)^P \subset LHS.$$

Let $\rho = (\rho_k)_{k=1}^s \in \bigoplus_{k=1}^s m(N)^q \mathscr{E}(r+N+1)^P$ – then for each $k = 1, \ldots, s$,

$$\begin{aligned}\rho_k(u,x,t) = \; & x^q \left(h_0^k(x) + s_1^k(t,u)h_1^k(x) + \ldots + s_{l-q-1}^k(t,u)h_{l-q-1}^k(x)\right. \\ & \left. + \; s_{l-q}^k(t,u)h_{l-q}^k(u,x,t)\right),\end{aligned}$$

where $s_i^k(t,u)$ is a homogeneous polynomial in $t, u_1, \ldots, u_r$ of degree i, and the h_i^k are smooth maps $\mathbb{R}^N \to \mathbb{R}^P$ for $i < l-q$, and the h_{l-q}^k are smooth maps $\mathbb{R}^{r+N+1} \to \mathbb{R}^P$.

The last term $s_{l-q}^k(t,u)h_{l-q}^k(u,x,t)$ is in $m(N)^q m(r+N+1)^{l-q}\mathscr{E}(r+N+1)^P$ as in the LHS of (70) by definition. Furthermore, LHS is closed under multiplication by t and u; hence it is enough to show that $(x_k^q h_0^k(x))_{k=1}^s \in LHS$ for any given h_0^k.

By (67) we can write $(x_k^q h_0^k(x))_{k=1}^s = tf(\xi) + wf(\zeta)$ for some elements

$$\xi \in \bigoplus_{k=1}^s m(N)^2\mathscr{E}(N)^N, \; \zeta \in m(P)^2\mathscr{E}(P)^P.$$

Thus

$$\bigsqcup_{k=1}^s x^q h_0^k(x) = \overbrace{(tf(\xi) - tF(\xi))}^{(*)} + \overbrace{(wf(\zeta) - wF(\zeta))}^{(**)} + tF(\xi) + wF(\zeta).$$

Now, $(*)$ and $(**)$ lie in $\bigoplus_{k=1}^s m(N)^l\mathscr{E}(r+N+1)^P$ and $\bigoplus_{k=1}^s m(N)^{l+1}\mathscr{E}(r+N+1)^P$, respectively, by Lemmas 68 and 69. It follows that $(*), (**) \subset LHS$ of (70), since $m(N)^l$ and $m(N)^{l+1}$ both sit inside $m(N)^q m(r+N+1)^{l-q}$.

The last two terms lie in LHS by definition. Hence, (70) holds.

Next, we prove (66) using (70) and a Nakayama-type of argument. We will use the following lemma by Mather:

Lemma 71. [Mat68a, Theorem 1.13] *Let $G\colon \bigsqcup_{k=1}^s(\mathbb{R}^n,0) \to (\mathbb{R}^p,0)$ be a smooth multigerm. Let A be a finitely generated $\mathscr{E}(p)$-module; let B and C be $\bigoplus_{k=1}^s \mathscr{E}(n)$-modules (with C finitely generated), $\beta\colon B \to C$ an $\bigoplus_{k=1}^s \mathscr{E}(n)$-module homomorphism; $\alpha\colon A \to C$ a homomorphism over $G^*\colon \mathscr{E}(p) \to \bigoplus_{k=1}^s \mathscr{E}(n)$. Denote by a the integer $\dim_{\mathbb{R}} A/m(p)A$.*

Then $\alpha(A)+\beta(B)+(G^*(m(p))+\bigoplus_{k=1}^s m(n)^{a+1})C = C$ implies $\alpha(A)+\beta(B) = C$.

Remark 72. Suppose that $D \subset C$ such that

$$\alpha(A) + \beta(B) + D = C \tag{73}$$

and

$$D \subset (G^*(m(p)) + \bigoplus_{k=1}^s m(n)^{a+1})C. \tag{74}$$

Then

$$\alpha(A) + \beta(B) = C. \tag{75}$$

We are going to use Lemma 71 with the substitutions:

$$
\begin{aligned}
&A := V\\
&B := \textstyle\bigoplus_{k=1}^{s} m(N)^2\mathscr{E}(r+N+1)^N\\
&C := tF(\textstyle\bigoplus_{k=1}^{s} m(N)^2\mathscr{E}(r+N+1)^N) + \textstyle\bigoplus_{k=1}^{s} m(N)^q\mathscr{E}(r+N+1)^P\\
&D := \textstyle\bigoplus_{k=1}^{s} m(N)^q m(r+N+1)^{l-q}\mathscr{E}(r+N+1)^P\\
&G := F\\
&\alpha := wF\\
&\beta := tF
\end{aligned}
$$

We check (73) and (74) – we see that (73) is exactly (70), and (74) is

$$
\begin{aligned}
&\textstyle\bigoplus_{k=1}^{s} m(N)^q m(r+N+1)^{l-q}\mathscr{E}(r+N+1)^P\\
&\subset \left(F^*(m(r+P+1)) + \textstyle\bigoplus_{k=1}^{s} m(r+N+1)^{a+1}\right)\cdot\\
&\cdot \left(tF(\textstyle\bigoplus_{k=1}^{s} m(N)^2\mathscr{E}(r+N+1)^N) + \textstyle\bigoplus_{k=1}^{s} m(N)^q\mathscr{E}(r+N+1)^P\right).
\end{aligned}
\tag{76}
$$

If we choose $l \geq a+q+1$, then (76) *must* hold. Thus (75) holds, which with our substitutions is exactly (66).

3.2.6. *Case II.* We solve the problem locally near a point in $\partial M \times [0,1]$, by reducing to the Case I.

Extend the map $F\colon \bigsqcup_s \mathbb{R}^r_+ \times \mathbb{R}^N \times \mathbb{R} \to \mathbb{R}^r_+ \times \mathbb{R}^P \times \mathbb{R}$ to a map $\tilde{F}\colon \bigsqcup_s \mathbb{R}^r \times \mathbb{R}^N \times \mathbb{R} \to \mathbb{R}^r \times \mathbb{R}^P \times \mathbb{R}$ still satisfying the conditions on F in the theorem. We need to show that

$$
\begin{aligned}
&\left(\left(\frac{\partial \tilde{F}^k_{y_1}}{\partial t}, \ldots, \frac{\partial \tilde{F}^k_{y_P}}{\partial t}\right)\right)_{k=1}^{s}\\
&\in m(N)^2 m(1)\cdot \textstyle\bigoplus_{k=1}^{s} \left\langle \left(\frac{\partial \tilde{F}^k_{y_1}}{\partial x_i}, \ldots, \frac{\partial \tilde{F}^k_{y_P}}{\partial x_i}\right) | i = 1, \ldots, N \right\rangle_{\mathscr{E}(r+N+1)}\\
&+ F^*\left(m(P)^2 m(1)\mathscr{E}(r+P+1)\right)^P,
\end{aligned}
\tag{77}
$$

where $m(1)$ refers to the ideal generated by the 1st local coordinate u_1 of M (and where ∂M is defined by $u_1 = 0$).

By (65), and because g and h have the same l-jets, LHS of (77) is contained in

$$
\underbrace{\bigoplus_{k=1}^{s} m(1)m(N)^{l+1}\mathscr{E}(r+N+1)^P}_{(*)}.
$$

Hence it is enough to show that $(*) \subset RHS$. Multiply (66), which we proved to be true in *Case I*, by $m(1)$, and we see that

$$
\bigoplus_{k=1}^{s} m(1)m(N)^{l+1}\mathscr{E}(r+N+1)^P
$$

is contained in

$$\bigoplus_{k=1}^{s} m(N)^2 m(1)\left\langle\left(\frac{\partial F^k_{y_1}}{\partial x_i},\ldots,\frac{\partial F^k_{y_P}}{\partial x_i}\right) | 1 \le i \le N\right\rangle_{\mathscr{E}(r+N+1)}$$
$$+ m(1)F^*\left(m(P)^2\mathscr{E}(r+P+1)\right)^P.$$

Hence it is enough to prove that

$$m(1)F^*\left(m(P)^2\mathscr{E}(r+P+1)\right)^P \subset F^*(m(P)^2 m(1)\mathscr{E}(r+P+1))^P.$$

This is clearly true, since a coordinate of the left hand side can be written in the form

$$u_1 \cdot (\eta \circ F)$$

for some $\eta \in \left(m(P)^2\mathscr{E}(r+P+1)\right)^P$, while a coordinate on the right hand side can be written in the form

$$(u_1 \cdot \eta) \circ F.$$

Hence the two sets are equal, and the proof of Proposition 57 is finished. □

We may now return to the proof of Theorem 56.

Denote by $\mathscr{A}^l$ the Lie group of l-jets of elements of $\mathscr{A}$ and set

$$\mathscr{A}^l_f = \{(z_1,\ldots,z_s,z_t) \in \mathscr{A}^l | z_t \circ j^l f \circ (z_1 \sqcup \ldots \sqcup z_s)^{-1} = j^l f\}.$$

For a sufficiently large $l \in \mathbb{N}$, the image of a maximal compact subgroup of $\mathscr{A}_f$ under j^l is a maximal compact subgroup of $\mathscr{A}^l_f$. Let G be a maximal compact subgroup of $\mathscr{A}_f$. By replacing f by a suitably chosen representative of its $\mathscr{A}$-equivalence class, we may assume that G acts linearly.

Given a manifold with boundary M, we must show that any smooth map

$$\alpha\colon \partial M \to \mathscr{A}_f/G$$

extends to a smooth map

$$\bar{\alpha}\colon M \to \mathscr{A}_f/G.$$

Lemma 78. *Suppose given a smooth map $\alpha\colon \partial M \to \mathscr{A}_f/G$. Then there exists a smooth lift $\tilde{\alpha}\colon \partial M \to \mathscr{A}_f$:*

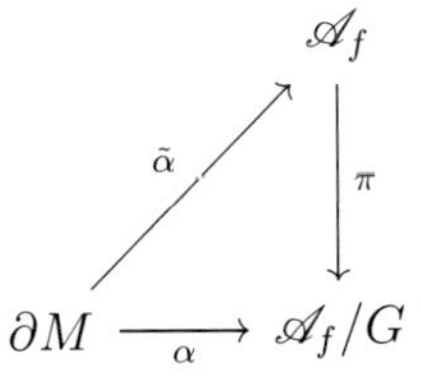

Proof. By the definition of a smooth map into $\mathscr{A}_f/G$, there exist a covering $\{U_i\}_{i\in I}$ of ∂M and smooth local lifts $\alpha_i\colon U_i \to \mathscr{A}_f$ such that $\pi \circ \alpha_i = \alpha | U_i$ for each $i \in I$. We need the lemma:

Lemma 79. *There is a G-equivariant map $h\colon \mathscr{A}^l_f \to G$.*

Proof. Since $\mathscr{A}_f^l$ is a Lie group, the quotient map $\mathscr{A}_f^l \to \mathscr{A}_f^l/G$ is a principal G-bundle projection, and since G is a maximal compact subgroup of $\mathscr{A}_f^l$, the quotient $\mathscr{A}_f^l/G$ is contractible. Hence the total space $\mathscr{A}_f^l$ is G-equivariantly equivalent to $G \times \mathscr{A}_f^l/G$ via a G-equivariant diffeomorphism $\varphi\colon \mathscr{A}_f^l \to G \times \mathscr{A}_f^l/G$. We set h to be the projection $\mathrm{pr}_G \circ \varphi$. □

We continue the proof of Lemma 78. Using Lemma 79, we can form a map

$$g_i\colon U_i \xrightarrow{\alpha_i} \mathscr{A}_f \xrightarrow{j^l} \mathscr{A}_f^l \xrightarrow{h} G$$

for each $i \in I$, and we define $\tilde{\alpha}\colon \partial M \to \mathscr{A}_f$ by setting $\tilde{\alpha}(x) = g_i(x)^{-1} \cdot \alpha_i(x)$ whenever $x \in U_i$. We show that α is well-defined:

Suppose that $x \in U_i \cap U_j$. There exists a unique $g \in G$ such that $\alpha_i(x) = g \cdot \alpha_j(x)$. Then

$$g_i(x) = h(j^l(\alpha_i(x))) = h(j^l(g \cdot \alpha_j(x))) = g \cdot h(j^l(\alpha_j(x))) = g \cdot g_j(x),$$

so

$$g_i(x)^{-1} \cdot \alpha_i(x) = (g \cdot g_j(x))^{-1} \cdot g \cdot \alpha_j(x) = g_j(x)^{-1} \cdot g^{-1} \cdot g \cdot \alpha_j(x) = g_j(x)^{-1} \cdot \alpha_j(x),$$

and it follows that $\tilde{\alpha}$ is well-defined. Furthermore, since $\tilde{\alpha}$ is locally smooth, it is smooth. This concludes the proof of Lemma 78. □

Corollary 80. *The proof of Lemma 78 holds also when $\mathscr{A}_f$ is replaced by all of $\mathscr{A}$, and if ∂M is replaced with any manifold (possibly with boundary) M. Hence, when G is a maximal compact subgroup of $\mathscr{A}$, any smooth map $M \to \mathscr{A}/G$ admits a smooth global lift $M \to \mathscr{A}$.* □

Remark 81. Jänich [Jän78] and Rimányi [Rim02] construct a lift $\tilde{\alpha}$ by using a section $\sigma\colon \mathscr{A}_f/G \to \mathscr{A}_f$ lifted from the section $\sigma^l\colon \mathscr{A}_f^l/G \to \mathscr{A}_f^l$, which exists because $\mathscr{A}_f^l/G$ is contractible. See the diagram (82) below. However, neither of them explain why it is possible to construct such a section σ, and we have not been able to come up with a proof.

(82)

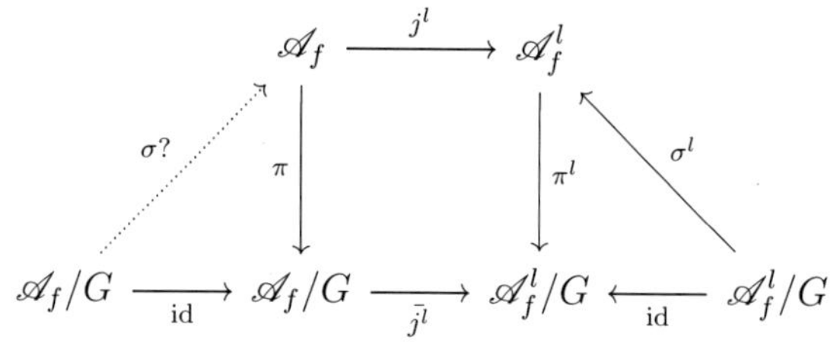

Consider the composition

$$\beta = \bar{j}^l \circ \alpha\colon \partial M \to \mathscr{A}_f/G \to \mathscr{A}_f^l/G.$$

Since $\mathscr{A}_f^l/G$ is contractible, we can construct an extension

$$\bar{\beta}\colon M \to \mathscr{A}_f^l/G,$$

and composing with the section σ^l we obtain a map

$$\gamma\colon M \to \mathscr{A}_f^l/G \to \mathscr{A}_f^l.$$

Lemma 83. *There exists a smooth map* $\delta\colon \partial M \to G$ *such that*

$$j^l \circ (\delta \cdot \tilde{\alpha})\colon \partial M \to \mathscr{A}_f^l$$

coincides with $\gamma|\partial M$.

Proof. Construct smooth maps

$$\begin{aligned} \delta_1\colon \partial M \to G, \quad \delta_1 = h \circ \gamma|\partial M, \\ \delta_2\colon \partial M \to G, \quad \delta_2 = h \circ \tilde{\alpha}, \end{aligned}$$

and set $\delta = \delta_1 \cdot \delta_2^{-1}$. Then

$$j^l \circ (\delta \cdot \tilde{\alpha}) = \gamma|\partial M$$

if and only if

(84) $$\varphi \circ j^l \circ (\delta \cdot \tilde{\alpha}) = \varphi \circ \gamma|\partial M,$$

and it is clear from the commutative diagram below that

$$\begin{aligned} \mathrm{pr}_{\mathscr{A}_f^l/G} \circ \varphi \circ j^l \circ (\delta \cdot \tilde{\alpha}) &= \pi^l \circ j^l \circ (\delta \cdot \tilde{\alpha}) \\ &= \pi^l \circ j^l \circ \tilde{\alpha} \\ &= \bar{\beta}|\partial M \\ &= \mathrm{pr}_{\mathscr{A}_f^l/G} \circ \varphi \circ \gamma|\partial M, \end{aligned}$$

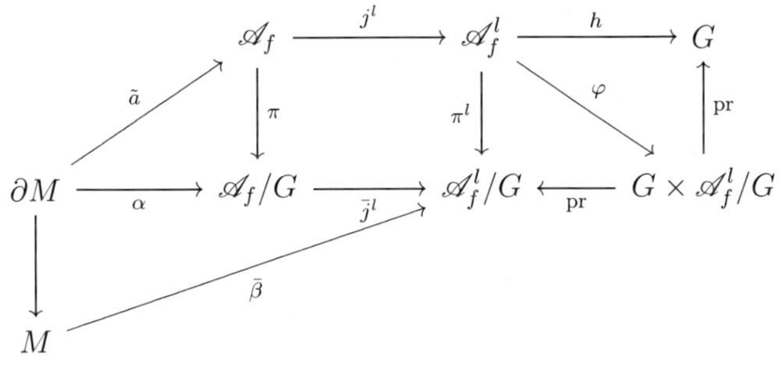

while

$$\begin{aligned} \mathrm{pr}_G \circ \varphi \circ j^l \circ (\delta \cdot \tilde{\alpha}) &= h \circ j^l \circ (\delta \cdot \tilde{\alpha}) \\ &= \delta \cdot (h \circ j^l \circ \tilde{\alpha}) \\ &= (h \circ \gamma|\partial M) \cdot (h \circ j^l \circ \tilde{\alpha})^{-1} \cdot (h \circ j^l \circ \tilde{\alpha}) \\ &= h \circ \gamma|\partial M \\ &= \mathrm{pr}_G \circ \varphi \circ \gamma|\partial M. \end{aligned}$$

Thus (84) is true, and this concludes the proof of Lemma 83. □

By Lemma 83 we see that replacing the old map $\tilde{\alpha}$ by $\delta \cdot \tilde{\alpha}$, we may assume that $j^l \circ \tilde{\alpha} = \gamma|\partial M$. This will enable us to construct a map $\bar{\tilde{\alpha}}\colon M \to \mathscr{A}$ which extends $\tilde{\alpha}$. Without the assumption $j^l \circ \tilde{\alpha} = \gamma|\partial M$ we risk – for instance, if $M = [0,1]$ – that $j^l(\tilde{\alpha}(0))$ and $j^l(\tilde{\alpha}(1))$ end up in different components of $\mathscr{A}_f^l$, in which case an extension of $\tilde{\alpha}$ is impossible. This, however, is not crucial in order to get the map into $\mathscr{A}_f/G$.

As a first step, we construct a map

$$\alpha'\colon M \to \mathscr{A}_{j^l f} = \{\phi \in \mathscr{A} | j^l\phi \in \mathscr{A}_f^l\},$$

extending $\tilde{\alpha}$, and in particular such that $\pi_{\mathscr{A}_{j^l f}} \circ \alpha'$ extends our map

$$\partial M \xrightarrow{\alpha} \mathscr{A}_f/G \to \mathscr{A}_{j^l f}/G,$$

where the second map is induced by the inclusion, and such that $j^l\alpha' = \gamma$.

In order to do this we must construct diffeomorphism germs at $M \times \{0\}$:

$$F_1 = \bigsqcup_{k=1}^s F_1^k\colon \left(\bigsqcup_s M \times \mathbb{R}^N, \bigsqcup_s M \times 0\right) \to \left(\bigsqcup_s M \times \mathbb{R}^N, \bigsqcup_s M \times 0\right)$$
$$F_2\colon \left(M \times \mathbb{R}^P, M \times 0\right) \to \left(M \times \mathbb{R}^P, M \times 0\right)$$

from given germs in $\partial M \times \bigsqcup_s \mathbb{R}^N$ and $\partial M \times \mathbb{R}^P$, and where the l-jets are given everywhere.

We go through the construction for one of the F_1^k; the proof for F_2 is similar. If we can find F_1^k locally, then we get a global solution by using a partition of unity to add the solutions together fiberwise. This gives a diffeomorphism germ since the l-jet, and thus in particular the differential DF_1^k, is fixed everywhere.

At points in the interior of M, we can just define the local F_1^k by taking the lth degree polynomial representative of the given jet. Near points in ∂M, we construct the component functions of the local F_1^k in the following way:

Given a polynomial P of degree l in the variables $x_1, \ldots, x_N$ with coefficients from the ring $\mathscr{E}(r)$ of smooth functions in r variables, and a smooth function $p_0\colon \mathbb{R}^{r-1+N} \to \mathbb{R}$ such that

$$j_x^l p_0 = P(0, u_2, \ldots, u_r, x_1, \ldots, x_N),$$

(here the coordinates of $\mathbb{R}^N$ are denoted by x_i and the local coordinates of M are denoted by u_j, where ∂M is given by $u_1 = 0$) we construct a smooth function $p\colon \mathbb{R}^{r+N} \to \mathbb{R}$ such that

$$p|\{u_1 = 0\} = p_0,$$

and

$$j_x^l p = P(u_1, \ldots, u_r, x_1, \ldots, x_N).$$

One function which satisfies all of the above, is the following:

$$p(u, x) = p_0(u_2, \ldots, u_r, x) - P(0, u_2, \ldots, u_r, x) + P(u_1, \ldots, u_r, x).$$

It follows that the map $\alpha'\colon M \to \mathscr{A}_{j^l f}$ exists, and it is represented by $F = (F_1^1, \ldots F_1^s, F_2)$. We use it to construct the wanted extension $\bar{\alpha}\colon M \to \mathscr{A}_f/G$:

We compare the maps

$$\phi_\nu\colon M \times \bigsqcup_a \mathbb{R}^n \to M \times \mathbb{R}^p \quad \nu = 1, 2$$

given by

$$\phi_1\colon (u,x) \mapsto (u, (\alpha'(u)\cdot f)(x))\,,$$
$$\phi_2\colon (u,x) \mapsto (u, f(x))\,.$$

These two maps coincide on $\partial M \times \bigsqcup_s \mathbb{R}^n$, and their l-jets coincide at each M-level, thus we can apply Proposition 57 to find a smooth map $\psi\colon M \to \mathscr{A}$ such that $\psi\cdot\phi_1 = \phi_2$, $j^1\psi = \mathrm{id}$ and such that $\psi|\partial M = \mathrm{id}$. Now the map

$$\bar{\bar{\alpha}}\colon M \to \mathscr{A}_f, \quad \bar{\bar{\alpha}} = \psi\cdot\alpha',$$

is an extension of $\tilde{\alpha}$ and $j^l\bar{\bar{\alpha}} = \gamma$ (because $j^1\psi = \mathrm{id}$). Most importantly, it defines an extension $\bar{\alpha} = \pi\circ\bar{\bar{\alpha}}\colon M \to \mathscr{A}_f/G$ of α. □

From the proof of Theorem 56, we see that the following corollary also holds:

Corollary 85. *Suppose that G is a maximal compact subgroup of $\mathscr{A}_f$. Given a smooth map $\alpha\colon \partial M \to \mathscr{A}_f$, we can find a smooth map $\gamma\colon \partial M \to G$ and a smooth map $\tilde{\alpha}\colon M \to \mathscr{A}_f$ such that $\tilde{\alpha}|\partial M = \gamma\cdot\alpha$.* □

3.3. **Factorization of $MC(\mathscr{A}_f)$ for multigerms.** The main goal of this section is to prove that maximal compact subgroups of $\mathscr{A}_f$ for multigerms f decompose into products of maximal compact subgroups of $\mathscr{A}_{g_i}$ for minimal representatives g_i of their associated monogerms:

Theorem 86. *We are given a ministable multigerm*

$$f = f_1 \sqcup \ldots \sqcup f_s\colon (\mathbb{R}^N,0)\sqcup\ldots\sqcup(\mathbb{R}^N,0) \to (\mathbb{R}^P,0),$$

where

$$f_i = \sigma_i\circ(g_i\times\mathrm{id})\colon \mathbb{R}^{n_i}\times\prod_{j=1,j\neq i}^{s}\mathbb{R}^{p_j} \xrightarrow{g_i\times\mathrm{id}} \mathbb{R}^{p_i}\times\prod_{j=1,j\neq i}^{s}\mathbb{R}^{p_j} \xrightarrow{\sigma_i} \prod_{j=1}^{s}\mathbb{R}^{p_j}, \tag{87}$$

where g_i is a ministable unfolding of a rank 0 representative $h_i\colon \mathbb{R}^{\tilde{n}_i} \to \mathbb{R}^{\tilde{p}_i}$ of f_i, and where $N = n_i + \sum_{j=1,j\neq i}^{s} p_j$ and $P = \sum_{j=1}^{s} p_j$.

Then the maximal compact subgroup factors as

$$MC(\mathscr{A}_f) \cong \prod_{j=1}^{s} MC(\mathscr{A}_{g_j}) \cong \prod_{j=1}^{s} MC(\mathscr{K}_{h_j}).$$

Remark 88. When $s=1$, this is [Wal80, Proposition 3.2].

Before proving this theorem, we need to study relations between maximal compact subgroups of $\mathscr{K}$- and $\mathscr{A}$-equivalences, the monogerm versions of which are well known. The following lemma is analogous to [Rim02, Theorem 1.1].

Lemma 89. *Let $f\colon \bigsqcup_s(\mathbb{R}^n,0) \to (\mathbb{R}^p,0)$ have rank 0 and let $F\colon \bigsqcup_s(\mathbb{R}^N,0) \to (\mathbb{R}^P,0)$ be a ministable unfolding of f. Suppose that $G < \mathscr{A}_F$ is a compact subgroup. Then there exists a compact subgroup of $\mathscr{K}_f$ which is isomorphic to G. In particular, $MC(\mathscr{A}_F) < MC(\mathscr{K}_f)$.*

Proof. Since G is compact, there exists $h \in \mathscr{A}$ such that $G_0 = hGh^{-1}$ is linear. If $F_0 = h \cdot F$, then $G_0 < \mathscr{A}_{F_0}$. We construct a map $f_0 \sim_{\mathscr{K}} f$ such that F_0 is a ministable unfolding of f_0.

Denote by $\Gamma(F_0)$ the graph of F_0. If we write $F_0 = \bigsqcup_{k=1}^{s} F_0^k$, then

$$\Gamma(F_0) = \bigsqcup_{k=1}^{s} \Gamma(F_0^k) = \bigsqcup_{k=1}^{s} \{(x, F_0^k(x)) | x \in \mathbb{R}^N\}.$$

Since G_0 is linear and $\Gamma(F_0)$ is G_0-invariant, the tangent space

$$T_0\Gamma(F_0) \subset (\mathbb{R}^N \sqcup \ldots \sqcup \mathbb{R}^N) \times \mathbb{R}^P$$

is also G_0-invariant.

Define subspaces

$$\begin{aligned} A^k &:= T_0(\Gamma(F_0^k)) \cap \left(\mathbb{R}^N \times \{0\}\right), \\ C^k &:= \mathrm{pr}_{\mathbb{R}^P}(T_0(\Gamma(F_0^k))), \end{aligned}$$

which are also G_0-invariant. Choose G_0-invariant complements B^k and D^k of A^k and C^k in $\mathbb{R}^N$ and $\mathbb{R}^P$, respectively. Then $A^k \cong \mathbb{R}^n$, $B^k \cong \mathbb{R}^r$, $C^k \cong \mathbb{R}^r$ and $D^k \cong \mathbb{R}^p$.

Denote by f_0 the map germ

$$\bigsqcup_{k=1}^{s} f_0^k \colon \bigsqcup_{k=1}^{s} (\mathrm{pr}_{D^k} \circ F_0^k)|A^k \colon \bigsqcup_{s} \mathbb{R}^n \to \mathbb{R}^p.$$

We shall prove next that through its action on $\bigsqcup_{k=1}^{s} \mathbb{R}^N \times \mathbb{R}^P$, G_0 is a subgroup of $\mathscr{K}_{f_0} = \mathscr{K}_{f_0^1} \times \cdots \times \mathscr{K}_{f_0^s}$

3.3.1. *F_0 is a ministable unfolding of f_0.* It follows from the definition of C^k that $\mathrm{pr}_{C^k} \circ F_0$ is a submersion for each k, so in particular the inclusion of D^k in $\mathbb{R}^P$ is transverse to the germ F_0^k. Furthermore, the diagram

$$\begin{array}{ccccc} A^k & \xrightarrow{f_0^k} & D^k & \longrightarrow & \mathbb{R}^p \\ \downarrow & & \downarrow & & \downarrow \\ A^k \times B^k & \xrightarrow[F_0^k]{} & D^k \times C^k & \longrightarrow & \mathbb{R}^p \times \mathbb{R}^r \end{array}$$

is trivially Cartesian for each k, and it follows that F_0 unfolds f_0. The unfolding F_0 is ministable because it is $\mathscr{A}$-equivalent to the ministable map F.

Note moreover that since F_0 and F are $\mathscr{A}$-equivalent, the maps f and f_0 must be $\mathscr{K}$-equivalent.

Project the group

$$G_0 < \prod_{k=1}^{s} GL(A^k) \times GL(B^k) \times GL(C^k) \times GL(D^k)$$

onto

$$\prod_{k=1}^{s} GL(A^k) \times GL(D^k); \tag{90}$$

then the resulting group lies in $\mathscr{A}_{f_0^1} \times \cdots \times \mathscr{A}_{f_0^s}$, so in particular it must lie in $\mathscr{K}_{f_0}$.

3.3.2. *The projection restricts to an injection on G_0.* Equivalently, the action of G_0 on the A^k and on D^k determines the action on the B^k and C^k. The actions on A^k, C^k and D^k determines that on B^k (being a germ, F_0 is level-preserving if the coordinates on B^k are appropriately chosen), thus it is enough to show that the action on C^k is determined by that on the A^k and D^k.

Recall from Chapter 1 that for a general ministable multigerm

$$\eta = \bigsqcup_{k=1}^{s} \eta_k \colon \bigsqcup_{k=1}^{s} (\mathbb{R}^N, 0) \to (\mathbb{R}^P, 0),$$

we define

$$\mathscr{N}_\eta = \theta_\eta / t\eta(\bigoplus_s \theta_{(\mathbb{R}^N,0)}) + \eta^*(m(P))\theta_\eta.$$

Each η_k is $\mathscr{A}$-equivalent to a germ

$$\tilde{\eta}_k \times \mathrm{id}_{\mathbb{R}^{d_k}},$$

so that $\tilde{\eta}_k \colon (\mathbb{R}^{n_k}, 0) \to (\mathbb{R}^{p_k}, 0)$ is ministable and has an isolated singularity.

We can decompose

$$\begin{aligned} &\theta_\eta \cong \textstyle\bigoplus_{k=1}^{s} \theta_{\eta_k} \\ &t\eta(\textstyle\bigoplus \theta_{(\mathbb{R}^n,0)}) \cong \textstyle\bigoplus_{k=1}^{s} t\eta_k(\theta_{(\mathbb{R}^n,0)}) \\ &\eta^*(m(p))\theta_\eta = \{f \circ \eta | f \colon (\mathbb{R}^p, 0) \to (\mathbb{R}, 0)\} \cdot \theta_\eta \cong \textstyle\bigoplus_{k=1}^{s} \eta_k^*(m(p))\theta_{\eta_k} \end{aligned}$$

and see that

$$\begin{aligned} \mathscr{N}_\eta &\cong \textstyle\bigoplus_{k=1}^{s} \theta_{\eta_k} / \bigoplus_{k=1}^{s} \left(t\eta_k(\theta_{(\mathbb{R}^n,0)}) + \eta_k^*(m(p))\theta_{\eta_k}\right) \\ &\cong \textstyle\bigoplus_{k=1}^{s} \theta_{\eta_k} / \left(t\eta_k(\theta_{(\mathbb{R}^n,0)}) + \eta_k^*(m(p))\theta_{\eta_k}\right) \\ &= \textstyle\bigoplus_{k=1}^{s} \mathscr{N}_{\eta_k} \\ &\cong \textstyle\bigoplus_{k=1}^{s} \mathscr{N}_{\tilde{\eta}_k}. \end{aligned}$$

In our situation, we write $F_0 = \bigsqcup_{k=1}^{s} F_0^k$ with $F_0^k = \sigma_k \circ \left(\tilde{F}_0^k \times \mathrm{id}_{\mathbb{R}^{P-p_k}}\right)$ and $\tilde{F}_0^k \colon \mathbb{R}^{n_k} \to \mathbb{R}^{p_k}$.

In the diagram

$$\begin{array}{ccc}
\mathbb{R}^P = \bigoplus_{k=1}^{s} \mathbb{R}^{p_k} = \bigoplus_{k=1}^{s} \left(\theta_{(\mathbb{R}^{p_k},0)}/m(p_k)\theta_{(\mathbb{R}^{p_k},0)}\right) & & \\
\Big\downarrow{\scriptstyle wF_0 = \bigoplus wF_0^k} & & \\
\mathscr{N}_{F_0} = \bigoplus_{k=1}^{s} \mathscr{N}_{F_0^k} & \xleftarrow{q_{F_0,f_0}} & \bigoplus_{k=1}^{s} \mathscr{N}_{f_0^k} = \mathscr{N}_{f_0}
\end{array}$$

there is a naturally defined G_0-action on each of the spaces, and both maps wF_0 and q_{F_0,f_0} are G_0-equivariant. Hence the action on $\mathscr{N}_{f_0}$ determines that on $\mathbb{R}^P$, and in particular that on the C^k. It follows that the projection (90) is injective on G_0.

Hence G_0 can be viewed as a compact subgroup of $\mathscr{K}_{f_0}$. Since the germs f and f_0 are $\mathscr{K}$-equivalent, the compact subgroups of $\mathscr{K}_f$ and $\mathscr{K}_{f_0}$ are conjugate in $\mathscr{K}$. There is thus a compact subgroup of $\mathscr{K}_f$ which is conjugate in $\mathscr{K}$, and hence isomorphic, to G_0, which again is isomorphic to G. This concludes the proof of Lemma 89. □

Proof of Theorem 86. Denote by h the rank 0 multigerm $\bigsqcup_{i=1}^{s} h_i$. Now we merely put the pieces together:

$$\begin{array}{ccc}
MC(\mathscr{A}_f) & \overset{\text{Lemma 89}}{<} & MC(\mathscr{K}_h) \\
 & \overset{\text{Proposition 45}}{\cong} & \prod_{i=1}^{s} MC(\mathscr{K}_{h_i}) \\
 & \overset{\text{Remark 88}}{\cong} & \prod_{i=1}^{s} MC(\mathscr{A}_{g_i}) \\
 & < & MC(\mathscr{A}_f),
\end{array}$$

where the last inequality is most easily seen to hold by considering the form (87) and taking each $((\psi_i, \phi_i)_{i=1}^{s}) \in \prod_{i=1}^{s} MC(\mathscr{A}_{g_i})$ to the element

$$\left(\bigsqcup_{i=1}^{s} \phi_1 \times \ldots \times \psi_i \times \ldots \times \phi_s, \phi_1 \times \ldots \times \phi_s\right).$$

But then the theorem holds. □

3.4. **Computing maximal compact subgroups.** We give some results facilitating computation of maximal compact subgroups of $\mathscr{K}_f$ for finitely determined rank 0 germs. Since we have shown that maximal compact subgroups of $\mathscr{K}_f$ for multigerms f can be decomposed as a product of maximal compact subgroups for the corresponding monogerms, it is enough to prove these results for monogerms.

Theorem 91. *Let $f\colon (\mathbb{R}^n, 0) \to (\mathbb{R}^p, 0)$ be finitely $\mathscr{K}$-determined, with $p < n$ and $T_0 f \equiv 0$. If $p > 1$, or if $p = 1$ and $j^2 f = 0$, then $MC(\mathscr{K}_f)$ is ≤ 1-dimensional, and if $p = 1$ then it is 0-dimensional.*

For $p = 1$, this is related to a theorem by P. Slodowy:

Theorem 92. [Slo78, Satz p. 169] *Let* $f\colon (\mathbb{R}^n, 0) \to (\mathbb{R}, 0)$ *be a germ such that* $j^2 f = 0$ *and* f *is finitely* $\mathscr{R}$*-determined. If a compact group* G *acts faithfully and linearly on* $\mathbb{R}^n$*, leaving* f *invariant, then* G *is zero-dimensional.* □

Remark 93. By [Wal81, Theorem 4.6.1], any $\mathscr{K}$-finitely determined function germ is $\mathscr{R}$-finitely determined.

Corollary 94. *Let* f *be as in Theorem 91 with* $p = 1$*, and let* $G < \mathscr{R}_f$ *be a compact subgroup. Then* G *is zero-dimensional.* □

Theorem 95. *Let* f *be as in Theorem 91 with* $p = 1$*, and let* $G < \mathscr{K}_f$ *be a compact subgroup. Then* G *is zero-dimensional.*

Proof. Changing f by a $\mathscr{K}$-equivalence, we may assume that G acts linearly by Lemma 48. Note furthermore that changing f by a $\mathscr{K}$-equivalence will not change the fact that $j^2 f = 0$.

Linear subgroups of $\mathscr{K}$ lie in $\mathscr{A}$ by Lemma 46; hence we can assume $G < GL_n \times GL_1 = GL_n \times \mathbb{R}^*$. The projections from $GL_n \times \mathbb{R}^*$ onto GL_n and $\mathbb{R}^*$ are continuous homomorphisms, and take G to compact subgroups $\tilde{G} < GL_n$ and $G_{\mathbb{R}} < \mathbb{R}^*$, respectively. Since $G_{\mathbb{R}}$ is a compact subgroup of $\mathbb{R}^*$, we must have $G_{\mathbb{R}} < \{\pm 1\}$.

Having this in mind, we see that $\tilde{G}$ splits into two parts, namely

$$\tilde{G} \cap \mathscr{R}_f \quad \text{and } H = \{g \in \tilde{G} | g \cdot f = -f\}.$$

The group $\tilde{G} \cap \mathscr{R}_f$ is finite by Theorem 92, but what about H?

Since $\tilde{G}$ is a Lie group, we must either have $\tilde{G}$ discrete, or $\tilde{G} \cap \mathscr{R}_f \subset \partial_{\tilde{G}} H$ with $\dim H \geq 1$. Suppose the latter. Then we can form a continuous path $\gamma\colon I \to \tilde{G}$ such that $\gamma(0) \in \mathscr{R}_f \cap \partial_{\tilde{G}} H$ and $\gamma(t) \in H$ for $t \neq 0$. Then we have $\gamma(t) \overset{t\to 0}{\to} \gamma(0)$, and for any given $x \in (\mathbb{R}^n, 0)$ we have $f(\gamma(t)(x)) \overset{t\to 0}{\to} f(\gamma(0)(x))$, since G is a matrix group. But by the definitions of H and $\mathscr{R}_f$, we have $f(\gamma(t)(x)) = -f(x)$ when $t \neq 0$, while $f(\gamma(0)(x)) = f(x)$, so unless $f(x) = 0$, this must be false. We have $f(x) \neq 0$ for x arbitrarily close to $0 \in \mathbb{R}^n$, and hence we cannot find such a path γ. But then $\tilde{G}$ must be discrete. Being a compact discrete set, $\tilde{G}$ is finite. □

For $p \geq 2$, C.T.C. Wall has proven an analogous result over the complex numbers:

Theorem 96. [Wal80, Theorem 3.3] *Let* $f\colon (\mathbb{C}^n, 0) \to (\mathbb{C}^p, 0)$ *have finite singularity type,* $1 < p < n$*, and* $T_0 f \equiv 0$*. Then*

$$\dim G_f \leq 1,$$

where G_f *is a maximal complex reductive subgroup of* $\mathscr{K}_f$*.* □

We shall pass from Wall's result to the corresponding claim over the real numbers from Theorem 91.

Proof of Theorem 91. Denote $G = MC(\mathscr{K}_f)$ for short. By Lemma 48, we may assume, up to a change of coordinates, that f is a polynomial and that G is linear. In particular,

$$G < \mathscr{K}_f \cap (GL_n \times GL_p) < \mathscr{A}_f$$

by Lemma 46.

There is a corresponding complex polynomial

$$f_{\mathbb{C}}\colon (\mathbb{C}^n, 0) \to (\mathbb{C}^p, 0)$$

with the same (real) coefficients as f. Then $f_{\mathbb{C}}$ is finitely $\mathscr{K}$-determined as well [Wal81, Proposition 1.7], hence has FST.

Viewing G as a subgroup of

$$GL(n, \mathbb{R}) \times GL(p, \mathbb{R}) < GL(n, \mathbb{C}) \times GL(p, \mathbb{C}),$$

we denote by $G_{\mathbb{C}}$ the Zariski closure of G in $GL(n, \mathbb{C}) \times GL(p, \mathbb{C})$. By Schwarz [Sch89, 2.2-2.6], the set $G_{\mathbb{C}}$ is a reductive complex algebraic subgroup of the algebraic group $GL(n, \mathbb{C}) \times GL(p, \mathbb{C})$, and if we write $\mathfrak{g}$ and $\mathfrak{g}_{\mathbb{C}}$ for the Lie algebras of G and $G_{\mathbb{C}}$, respectively, then $\mathfrak{g}_{\mathbb{C}} = \mathfrak{g} + i\mathfrak{g}$.

We argue that $G_{\mathbb{C}} < \mathscr{A}_{f_{\mathbb{C}}}$, which will prove that $\dim_{\mathbb{C}} G_{\mathbb{C}} \leq 1$ by Wall's theorem (that is, Theorem 96).

The action of $G_{\mathbb{C}}$ on $(\mathbb{C}^n, 0) \times (\mathbb{C}^p, 0)$ is algebraic, and hence Zariski continuous. Viewing G as a subset of $G_{\mathbb{C}}$ with the induced Zariski topology, the maps

$$\begin{aligned} \Phi\colon G \to \mathbb{C}^p, &\quad g \mapsto (g \cdot f)_{\mathbb{C}}(z) \\ \Phi_{\mathbb{C}}\colon G_{\mathbb{C}} \to \mathbb{C}^p, &\quad g_{\mathbb{C}} \mapsto (g_{\mathbb{C}} \cdot f_{\mathbb{C}})(z) \end{aligned}$$

are Zariski continuous for any fixed $z \in \mathbb{C}^n$, and $\Phi_{\mathbb{C}}$ is a continuous extension of Φ.

The map Φ is constant, because $G < \mathscr{A}_f$ and hence $g \cdot f = f$ for all $g \in G$. But G is Zariski dense in $G_{\mathbb{C}}$ by [Sch89], and points are closed in the Zariski topology on $\mathbb{C}^p$; hence $\Phi_{\mathbb{C}}$ must be constant as well. Since this holds for all $z \in \mathbb{C}^n$, it follows that $G_{\mathbb{C}} < \mathscr{A}_{f_{\mathbb{C}}}$, and $\dim G_{\mathbb{C}} \leq 1$ by Theorem 96.

Then $\dim_{\mathbb{C}} \mathfrak{g}_{\mathbb{C}} \leq 1$, and since $\mathfrak{g}_{\mathbb{C}} = \mathfrak{g} + i\mathfrak{g}$, we must have $\dim_{\mathbb{R}} \mathfrak{g} \leq 1$, and in particular $\dim_{\mathbb{R}} G \leq 1$. □

3.4.1. *Computation of maximal compact subgroups.* Using the results above, we are able to perform some actual computations.

Theorem 97. *Let $f\colon (\mathbb{R}^2, 0) \to (\mathbb{R}, 0)$ be one of the map germs*

$$\begin{aligned} &a) \quad (x, y) \mapsto y^3 + \lambda y x^{2p} + x^{3p}, \\ &b) \quad (x, y) \mapsto x(y^3 + \lambda y x^{2p} + x^{3p}), \end{aligned} \tag{98}$$

with $p > 1$ and $\lambda \neq 0$ (that is, f is a normal form up to $\mathscr{K}$-equivalence for the $E_{p,0}(\lambda)$- or $Z_{p,0}(\lambda)$-singularity, both of which we shall meet in the next chapter).

These are weighted homogeneous polynomials as defined in Chapter 2.4, and f is $\mathbb{R}^*$-equivariant. But then $\mathbb{R}^* < \mathscr{A}_f < \mathscr{K}_f$. We claim that $\{\pm 1\} < \mathbb{R}^*$ is a maximal compact subgroup of $\mathscr{K}_f$.

In order to simplify calculations, we decide to work not with f, but with the $\mathscr{R}$-equivalent function germs $\tilde{f}\colon (\mathbb{R}^2, 0) \to (\mathbb{R}, 0)$ given by

$$\tilde{f}(x, y) = f(y, x)\colon \mathbb{R}^2 \to \mathbb{R},$$

that is

$$\begin{array}{ll} a) & (x,y) \mapsto x^3 + \lambda x y^{2p} + y^{3p}, \\ b) & (x,y) \mapsto y(x^3 + \lambda x y^{2p} + y^{3p}). \end{array}$$

The maximal compact subgroups of $\mathscr{K}_f$ and $\mathscr{K}_{\tilde{f}}$ are conjugate in $\mathscr{K}$, so they are isomorphic.

Let $G < \mathscr{K}_{\tilde{f}}$ be a maximal compact subgroup. Suppose that $(l,(h_1,h_2)) \in G$, which acts on $\tilde{f}$ by

$$(l,(h_1,h_2)) \cdot \tilde{f}(x,y) = l(x,y) \cdot \tilde{f}(h_1(x,y), h_2(x,y)).$$

We note that l is completely determined by (h_1,h_2); in other words G is determined by its action on the source space, and the projection

$$p\colon \mathscr{K} = \mathscr{C} \rtimes \mathscr{R} \to \mathscr{R}, \quad (l,(h_1,h_2)) \mapsto (h_1,h_2),$$

restricts to an injection on $\mathscr{K}_f$. We start out by investigating the 1-jet of $h = (h_1,h_2)$, denoted

$$j^1 h = \begin{bmatrix} \alpha & \beta \\ \gamma & \delta \end{bmatrix}.$$

Proposition 99. *In this situation,*

a) $\beta = 0$,

b) $\alpha, \delta \in \{\pm 1\}$,

c) $j^1 h = \begin{bmatrix} 1 & 0 \\ \gamma & 1 \end{bmatrix}$ *can only hold if* $\gamma = 0$,

d) $j^1 h = \begin{bmatrix} -1 & 0 \\ \gamma & 1 \end{bmatrix}$ *never holds,*

e) $j^1 h = \begin{bmatrix} -1 & 0 \\ \gamma & -1 \end{bmatrix}$ *can only hold if* p *is odd,*

f) $j^1 h = \begin{bmatrix} 1 & 0 \\ \gamma & -1 \end{bmatrix}$ *can only hold if* p *is even.*

Proof.

a) Consider

$$l(x,y) \cdot \tilde{f}(h_1(x,y), h_2(x,y)) = \begin{cases} a) & l(h_1^3 + \lambda h_2 h_1^{2p} + h_2^{3p})(x,y), \\ b) & l h_2 (h_1^3 + \lambda h_2 h_1^{2p} + h_2^{3p})(x,y). \end{cases}$$

There is a term of weight $< 3p$ if and only if h_1 has a term of weight $< p$, but even then the terms h_1^3 and $\lambda h_1 h_2^{2p}$ cannot add up to the same weight. Thus, h_1 has no term of weight $< p$; and its 1-jet $j^1 h_1$ is just αx for some $\alpha \in \mathbb{R}$, and hence, $\beta = 0$.

b) By Theorem 91, G is finite, so $j^1 G$ is finite, and $(j^1 h)^n = 1$ for some $n \in \mathbb{N}$. But

$$(j^1 h)^n = \begin{bmatrix} \alpha^n & 0 \\ * & \delta^n \end{bmatrix},$$

so we must have $\alpha, \delta \in \{\pm 1\}$.

c) As in *b)*, we use the fact that $(j^1h)^n = 1$ for some n, and note that

$$(j^1h)^n = \begin{bmatrix} 1 & 0 \\ n\gamma & 1 \end{bmatrix},$$

so we must have $\gamma = 0$.

d) Suppose that we have $j^1h = \begin{bmatrix} -1 & 0 \\ \gamma & 1 \end{bmatrix}$; then

$$\begin{aligned} h_1(x,y) &= -x + ay^p + \tilde{h}_1(x,y), && j^1\tilde{h}_1 = 0, a \in \mathbb{R} \text{ and } j^p\tilde{h}_1 \\ & && \text{is a function of } x. \\ h_2(x,y) &= \gamma x + y + \tilde{h}_2(x,y), && j^1\tilde{h}_2 = 0. \end{aligned}$$

Then $\tilde{f}(x,y)$ equals, up to multiplication with $l(x,y)^{-1}$,

$$\begin{aligned}
&\tilde{f}(h_1(x,y), h_2(x,y)) \\
&= \begin{cases} a) & (h_1^3 + \lambda h_1 h_2^{2p} + h_2^{3p})(x,y) \\ b) & h_2(h_1^3 + \lambda h_1 h_2^{2p} + h_2^{3p})(x,y) \end{cases} \\
&= \begin{cases} a) & (-x + ay^p)^3 + \lambda(-x + ay^p)y^{2p} + y^{3p} + H.O.T \\ b) & y\left((-x + ay^p)^3 + \lambda(-x + ay^p)y^{2p} + y^{3p} + H.O.T\right) \end{cases} \\
&= \begin{cases} a) & (-x)^3 + 3(-x)^2ay^p + 3(-x)(ay^p)^2 + (ay^p)^3 - \lambda xy^{2p} \\ & + a\lambda y^{3p} + y^{3p} + H.O.T \\ b) & y\left((-x)^3 + 3(-x)^2ay^p + 3(-x)(ay^p)^2 + (ay^p)^3 - \lambda xy^{2p}\right. \\ & \left. + a\lambda y^{3p} + y^{3p} + H.O.T\right) \end{cases} \\
&= \begin{cases} a) & -x^3 + 3ax^2y^p - 3a^2xy^{2p} + a^3y^{3p} - \lambda xy^{2p} + a\lambda y^{3p} + y^{3p} \\ & + H.O.T \\ b) & y\left(-x^3 + 3ax^2y^p - 3a^2xy^{2p} + a^3y^{3p} - \lambda xy^{2p} + a\lambda y^{3p} + y^{3p}\right. \\ & \left. + H.O.T\right) \end{cases} \\
&= \begin{cases} a) & -x^3 + 3ax^2y^p - (3a + \lambda)xy^{2p} + (a^3 + a\lambda + 1)y^{3p} + H.O.T \\ b) & y\left(-x^3 + 3ax^2y^p - (3a + \lambda)xy^{2p} + (a^3 + a\lambda + 1)y^{3p}\right. \\ & \left. + H.O.T\right) \end{cases}
\end{aligned}$$

which implies $3a = 0$, $3a + \lambda = \lambda$ and $a^3 + \lambda a + 1 = -1$, which is impossible.

e) Suppose that we have $j^1h = \begin{bmatrix} -1 & 0 \\ \gamma & -1 \end{bmatrix}$. Then we can write

$$\begin{aligned} h_1(x,y) &= -x + ay^p + \tilde{h}_1(x,y) && j^1\tilde{h}_1 = 0 \text{ and } j^p\tilde{h}_1 \text{ is a function} \\ & && \text{of } x, a \in \mathbb{R}. \\ h_2(x,y) &= \gamma x - y + \tilde{h}_2(x,y), && j^1\tilde{h}_2 = 0. \end{aligned}$$

We write the proof for the case a); the proof b) follows the same recipe just like in d). Then $\tilde{f}(x,y)$ equals, up to multiplication with $l(x,y)^{-1}$,

$$
\begin{aligned}
&\tilde{f}(h_1(x,y),h_2(x,y))\\
&=(h_1^3+\lambda h_1 h_2^{2p}+h_2^{3p})(x,y)\\
&=(-x+ay^p)^3+\lambda(-x+ay^p)(-y)^{2p}+(-y)^{3p}+H.O.T\\
&=-x^3+3ax^2y^p-3a^2xy^{2p}+a^3y^{3p}-\lambda xy^{2p}+a\lambda y^{3p}+(-1)^{3p}y^{3p}\\
&\quad+H.O.T.\\
&=-x^3+(3a)x^2y^p-(3a^2+\lambda)xy^{2p}+(a^3+a\lambda+(-1)^{3p})y^{3p}+H.O.T.
\end{aligned}
$$

which implies $3a=0$, $3a^2+\lambda=\lambda$ and $a^3+a\lambda+(-1)^{3p}=-1$, which only holds if $a=0$ and p is odd.

f) Suppose that we have $j^1h=\begin{bmatrix}1 & 0\\ \gamma & -1\end{bmatrix}$. Then we can write

$$
\begin{aligned}
h_1(x,y)&=x+ay^p+\tilde{h}_1(x,y) \quad j^1\tilde{h}_1=0 \text{ and } j^p\tilde{h}_1 \text{ is a function of } x,a\in\mathbb{R}.\\
h_2(x,y)&=\gamma x-y+\tilde{h}_2(x,y),\quad j^1\tilde{h}_2=0.
\end{aligned}
$$

Again, we write out the calculation for the germ a). As in d), the calculation for the germ b) is completely the same. Then $\tilde{f}(x,y)$ equals, up to multiplication with $l(x,y)^{-1}$,

$$
\begin{aligned}
&\tilde{f}(h_1(x,y),h_2(x,y))\\
&=(h_1^3+\lambda h_1 h_2^{2p}+h_2^{3p})(x,y)\\
&=(x+ay^p)^3+\lambda(x+ay^p)(-y)^{2p}+(-y)^{3p}+H.O.T.\\
&=x^3+3ax^2y^p+3a^2xy^{2p}+a^3y^{3p}+\lambda xy^{2p}+a\lambda y^{3p}+(-1)^{3p}y^{3p}\\
&\quad+H.O.T.\\
&=x^3+(3a)x^2y^p+(3a^2+\lambda)xy^{2p}+(a^3+a\lambda+(-1)^{3p})y^{3p}+H.O.T.
\end{aligned}
$$

which implies $3a=0$, $3a^2+\lambda=\lambda$ and $a^3+a\lambda+(-1)^{3p}=1$, which only holds if $a=0$ and p is even.

And this ends the proof of the proposition. □

We have now showed that G consists of the identity and

i) elements $(l,(h_1,h_2))$ such that $j^1h=\begin{bmatrix}-1 & 0\\ \gamma & -1\end{bmatrix}$ if p is odd, or

ii) elements $(l,(h_1,h_2))$ such that $j^1h=\begin{bmatrix}1 & 0\\ \gamma & -1\end{bmatrix}$ if p is even.

Proposition 100. *Let G be a maximal compact subgroup of $\mathscr{K}_{\tilde{f}}$, and let the triple $(l,(h_1,h_2))$ be an element of G such that*

i) $j^1h=\begin{bmatrix}-1 & 0\\ \gamma & -1\end{bmatrix}$, *if p is odd, and*

ii) $j^1h=\begin{bmatrix}1 & 0\\ \gamma & -1\end{bmatrix}$ *if p is even,* *for a fixed number* $\gamma \in \mathbb{R}$*. Then*

i) $\gamma = 0$*, or*

ii) *there is no other* $\gamma' \in \mathbb{R}$ *such that* G *also contains an element* $(l', (h'_1, h'_2))$ *with* $j^1h' = \begin{bmatrix} 1 & 0 \\ \gamma' & -1 \end{bmatrix}$.

As a consequence, $G \cong \mathbb{Z}_2$*.*

Proof.

i) Now j^1G is finite and contains

$$\begin{bmatrix} -1 & 0 \\ \gamma' & -1 \end{bmatrix} \begin{bmatrix} -1 & 0 \\ \gamma' & -1 \end{bmatrix} = \begin{bmatrix} 1 & 0 \\ -2\gamma' & 1 \end{bmatrix},$$

and hence also $\begin{bmatrix} 1 & 0 \\ -2\gamma' & 1 \end{bmatrix}^n = \begin{bmatrix} 1 & 0 \\ -2n\gamma' & 1 \end{bmatrix}$ for all $n \in \mathbb{N}$. This is only possible of $\gamma = 0$.

ii) If (l, h) and (l', h') are both in G, then the 1-jets of h and h' are both in j^1G. But then again, so is their product:

$$\begin{bmatrix} 1 & 0 \\ \gamma & -1 \end{bmatrix} \begin{bmatrix} 1 & 0 \\ \gamma' & -1 \end{bmatrix} = \begin{bmatrix} 1 & 0 \\ \gamma - \gamma' & 1 \end{bmatrix}$$

Set $\delta = \gamma - \gamma'$; now also

$$\begin{bmatrix} 1 & 0 \\ \delta & 1 \end{bmatrix}^n = \begin{bmatrix} 1 & 0 \\ n\delta & 1 \end{bmatrix}$$

is contained in j^1G for each $n \in \mathbb{N}$. Thus the only way that j^1G can be finite is if $\delta = 0$; that is $\gamma = \gamma'$.

This proves the proposition. □

Corollary 101. *We return to the original polynomial* f*; from the previous proposition we can conclude that* $MC(\mathscr{K}_f) \cong \mathbb{Z}_2$*.*

Proof. We know that f and $\tilde{f}$ are $\mathscr{K}$-equivalent, hence $MC(\mathscr{K}_f)$ and $MC(\mathscr{K}_{\tilde{f}})$ are isomorphic, hence $MC(\mathscr{K}_f)$ is isomorphic to $\mathbb{Z}_2$. □

Proof of Theorem 97. We have noted that $\mathbb{R}^*$ is a subgroup of $\mathscr{K}_f$; hence the maximal compact subgroup of $\mathbb{R}^*$ will be a compact subgroup of $\mathscr{K}_f$, and it will be contained in a maximal compact subgroup of $\mathscr{K}_f$. But the maximal compact subgroup of $\mathbb{R}^*$ is $\{\pm 1\}$, which is isomorphic to $\mathbb{Z}_2$. Hence the maximal compact subgroup of $\mathscr{K}_f$ containing $\{\pm 1\}$ is just $\{\pm 1\}$ itself. □

3.4.2. *Consequences for* $\mathscr{A}_F$ *for a stable multigerm* F*.* The previous results allow us to really compute maximal compact subgroups of $\mathscr{A}_F$, for stable multigerms F, and we can deduce some very nice properties of $\mathscr{A}_F$ itself.

Corollary 102. *Suppose that* $F\colon \bigsqcup_s(\mathbb{R}^N,0)\to(\mathbb{R}^P,0)$ *is a ministable multigerm, unfolding the rank* 0 *multigerm*

$$f=\bigsqcup_{i=1}^{s} f_i\colon \bigsqcup_{i=1}^{s}(\mathbb{R}^{n_i},0)\to(\mathbb{R},0),$$

where the f_i *are of the form (98). By Theorem 86 and Theorem 91, we have*

$$MC(\mathscr{A}_F)\cong\prod_{i=1}^{s} MC(\mathscr{K}_{f_i})\cong\underbrace{\{\pm1\}\times\ldots\times\{\pm1\}}_{s\text{ times}},$$

and in particular, $MC(\mathscr{A}_F)$ *is finite.* □

4. Construction of E-tame retractions in examples

4.1. E-singularities. We study the singularities $E_{p,0}$ with normal forms

$$(x,y)\mapsto y^3+yx^{2p}W_{p-1}(x)+x^{3p} \tag{103}$$

and

$$(x,y)\mapsto y^3+yx^{2p}+x^{3p}W_{p-1}(x) \tag{104}$$

where

$$W_{p-1}(x)=w_0+w_1x+\ldots+w_{p-2}x^{p-2}.$$

We require that $4w_0^3+27\neq0$ or $4+27w_0^2\neq0$, respectively, in order to have finitely $\mathscr{K}$-determined germs.

Different values of w_0 or $r=\min\{i>0|w_i\neq0\}$ in the maps (103) and (104) give $\mathscr{K}$-distinct germs, and the $\mathscr{K}$-equivalence classes of singularities of type $E_{p,0}$ are parametrized by $w_0\in\mathbb{R}$ and $r=1,\ldots,p-1$ through the representatives

$$y^3+yx^{2p}(w_0+x^r)+x^{3p}, \tag{105}$$

and

$$y^3+yx^{2p}+x^{3p}(w_0+x^r).$$

These representatives do not only describe a subset of the set of maps, but descend to representatives of classes in the set of germs $(\mathbb{R}^m,0)\to(\mathbb{R}^{m-1},0)$, and jet space $J^k(m,m-1)$ modulo $\mathscr{E}\mathscr{K}$-equivalence, where $\mathscr{E}\mathscr{K}$-equivalence refers to extended $\mathscr{K}$-equivalence, as defined in Chapter 1. Taking $\mathscr{K}$-saturations, we obtain $\mathscr{K}$-invariant subsets of the space of germs, and of jet space.

In fact, the cases of (103) and (104) where $r=p-1$ are equivalent to the cases with $W_{p-1}(x)=w_0$, and we obtain the weighted homogeneous normal forms

$$y^3+w_0yx^{2p}+x^{3p} \tag{106}$$

and

$$y^3+yx^{2p}+w_0x^{3p} \tag{107}$$

They admit standard ministable weighted homogeneous unfoldings $F_{(p,0)}\colon \mathbb{R}^2 \times \mathbb{R}^{6p-3} \to \mathbb{R} \times \mathbb{R}^{6p-3}$ given by

$$(108) \qquad (x, y, \underline{u}, \underline{v}) \mapsto (y^3 + w_0 y x^{2p} + x^{3p} + \sum_{i=1}^{3p-2} u_i x^i + y \sum_{j=0}^{3p-2} v_j x^j, \underline{u}, \underline{v}),$$

and

$$(109) \qquad (x, y, \underline{u}, \underline{v}) \mapsto (y^3 + y x^{2p} + w_0 x^{3p} + \sum_{i=1}^{3p-2} u_i x^i + y \sum_{j=0}^{3p-2} v_j x^j, \underline{u}, \underline{v}),$$

where the weighted homogeneity is obtained by assigning appropriate weights to the u_i and v_j variables. Note that all the normal forms given in (103) and (104) appear as germs at appropriately chosen points, on appropriately chosen submanifolds of source and target, in (108) and (109).

We can prove that all the cases of (107) with $w_0 \neq 0$ are $\mathscr{A}$-equivalent to some case of (106) with another $w_0 \neq 0$; more precisely, we show that a germ of the form

$$(110) \qquad y^3 + a y x^{2p} + b x^{3p},$$

with $a, b \neq 0$, can be taken to the form (106) by a smooth change of variables. Transform $x \rightsquigarrow \frac{1}{b^{\frac{1}{3p}}} x$; then (110) becomes

$$\tilde{y}^3 + a y (\frac{1}{b^{\frac{1}{3p}}})^{2p} x^{2p} + b (\frac{1}{b^{\frac{1}{3p}}})^{3p} x^{3p} = y^3 + a b^{-\frac{2}{3}} y x^{2p} + x^{3p},$$

and it follows that (110), and in particular (107), is $\mathscr{A}$-equivalent to a germ of the type (106).

In a moment, we shall justify treating only the cases with $w_0 \neq 0$; for now we just accept the decision to leave the $w_0 = 0$ germs out, and conclude that then we only need to work with the case (106). Upon simplifying our normal form to (106), we simplify the writing a little more by renaming $w_0 \rightsquigarrow \lambda$, giving the normal form

$$(111) \qquad f_{(p,0)}(\lambda)(x, y) = y^3 + \lambda y x^{2p} + x^{3p}.$$

It turns out that the weights assigned to the u_i are always positive, while $p-1$ of the weights assigned to the v_j will be non-positive, and we consider the restriction

$$F^+_{(p,0)} = F_{(p,0)} | \{(x, y, \underline{u}, \underline{v}) | v_j = 0 \text{ if } wt(v_j) \leq 0\} \colon \mathbb{R}^{5p} \to \mathbb{R}^{5p-1}.$$

We know that $F^+_{(p,0)}$ is not smoothly stable, but by constructing a retraction $(r, s)\colon F_{(p,0)} \to F^+_{(p,0)}$ which is E-tame, we can show that $F^+_{(p,0)}$ is topologically stable (in fact, it is topologically ministable), and that the topological type of smoothly stable $E_{p,0}(\lambda)$ germs is constant for $\lambda \neq 0$. It has been shown [dPW04, Theorem 2.3] that the germs corresponding to $w_0 = 0$ in (106) and (107) admit some deformations which are different from those which we find for $w_0 \neq 0$; hence we cannot find our retraction unless we also require that $w_0 \neq 0$ – and our decision to leave the $w_0 = 0$ case out is thus justified and necessary.

We shall see that by finding E-tame retraction germs for the weighted homogeneous case (111), we actually obtain E-tame retraction germs for all the cases (103) and (104) with $w_0 \neq 0$ by using the $\mathbb{R}^+$-action, see Section 4.3.

Topological triviality of a weighed homogeneous, smoothly stable unfolding over its positively weighted part was shown for $E_{3,0}(\lambda)$ $(\lambda \neq 0)$ by Damon and Galligo [DG93]; however, their trivialization does not prove topological stability. Topological triviality and stability for $E_{2,0}(\lambda)$ $(\lambda \neq 0)$ was shown by Looijenga [Loo77] a long time ago. Tame retractions for $E_{2,0}(\lambda)$ and $E_{3,0}(\lambda)$ are constructed in the book by du Plessis and Wall [dPW95], and we will construct E-tame retractions for these as part of our construction for $E_{4,0}(\lambda)$ $(\lambda \neq 0)$.

4.1.1. *Presentation of E-singularities in jet space, source and target.* We start by investigating the subset of jet space defined by $E_{p,0}$. Let $p \in \mathbb{N}$ and $m \in \mathbb{N}$, $m \geq 2$; by abuse of notation we denote by $E_{p,0}$ the set of jets in $J^k(m, m-1)$ which are $\mathscr{E}\mathscr{K}^k$-equivalent to some jet of the form (103) or (104) with $4w_0^2 + 27 \neq 0$ or $4 + 27w_0^3 \neq 0$. We denote by $E_{p,0}(*)$ the subset of $E_{p,0}$ consisting of jets with $w_0 \neq 0$; i.e. defined by (111) with $\lambda \neq 0$.

Proposition 112. *For sufficiently large $k \in \mathbb{N}$, the subset $E_{p,0}(*)$ [$E_{p,0}$] is a smooth submanifold of $J^k(m, m-1)$.*

Proof. We prove the proposition utilizing techniques from [dP99], using a stable unfolding with unfolding variables from jet space. Our proof will be for $E_{p,0}(*)$, but the same proof applies to $E_{p,0}$ as well.

It is enough to prove that the space-germ $(E_{p,0}(*), w)$ is a smooth submanifold for any $w \in E_{p,0}(*)$.

Define $W^k(n,p) = \{z \in J^k(n,p) | d_e(z, \mathscr{K}) \geq k\}$. Then every jet in $J^k(n,p) \setminus W^k(n,p)$ is $\mathscr{K}$-sufficient by the determinacy theorem; see Proposition 6.

For simplicity we denote $A = J^k(m, m-1) \setminus W^k(m, m-1)$, and define a map

$$\Phi \colon \mathbb{R}^m \times A \to \mathbb{R}^{m-1} \times A$$

by setting

$$\Phi(x, z) = (\tilde{z}(x), z),$$

where $\tilde{z}$ is the deg $\leq k$ polynomial representing z.

Lemma 113. *The map Φ is smooth.*

Proof. Identify $J^k(m, m-1)$ with the $(m-1)^{\text{th}}$ power of the space of coefficients of monomials of degree $\leq k$ in m variables; now $J^k(m, m-1)$ has the Euclidean topology. In these coordinates, Φ is just a polynomial map and it must be smooth. □

Lemma 114. *Write $\hat{\Phi}_z$ for the germ of Φ at $(0, z)$. Then $\hat{\Phi}_z$ is a stable unfolding of $\tilde{z}$, so in particular, $\hat{\Phi}_z$ is a stable germ which is $\mathscr{E}\mathscr{K}$-equivalent to $\tilde{z}$.*

Proof. See [dP99, p. 3]. □

By Lemma 114, the jet of $\hat{\Phi}_z$ belongs to $E_{p,0}(*)$ if and only if z belongs to $E_{p,0}(*)$.

Return to our $w \in E_{p,0}(*)$; form the standard ministable unfolding $F_{(p,0)}$ of $\tilde{w}$, and note that $\hat{\Phi}_w$ is $\mathscr{A}$-equivalent to $\hat{F}_{(p,0)} \times \mathrm{id}_{(\mathbb{R}^d,0)}$ for some $d \in \mathbb{N}_0$. Using the $\mathscr{A}$-equivalence, we see that the germ of $\{0\} \times E_{p,0}(*)$ at $(0,w)$ in $\mathbb{R}^{m-1} \times A$ is diffeomorphic to the space-germ

$$\underbrace{\text{"Presentation of } E_{p,0}(*) \text{ by } F_{(p,0)}\text{"} \times (\mathbb{R}^d, 0)}_{(*)} \bigcap \{0_{m-1}\} \times J^k(m, m-1)$$

at $(0,0)$. The set $(*)$ is just $0_{\mathbb{R}^{5p}} \times \mathbb{R}^{p-1} \times \mathbb{R}^d$, hence is a smooth submanifold. Arguing as in [dP99, Proposition 1.1], we shall show that $(*)$ is transverse to $\{0_{m-1}\} \times J^k(m, m-1)$, which is enough to prove the proposition.

The set $(*)$ corresponds to the subset of $\mathbb{R}^{m-1} \times A$ where $E_{p,0}(*)$-singularities are presented by Φ. For any polynomial map

$$p\colon (\mathbb{R}^n, 0) \to (\mathbb{R}^p, 0)$$

of degree l we see that for all $y \in \mathbb{R}^n$, the map

$$p_y\colon (\mathbb{R}^n, 0) \to (\mathbb{R}^p, 0), \quad p_y(x) = p(x-y) - p(-y)$$

is a polynomial of degree $\leq l$ whose germ at y is $\mathscr{A}$-equivalent to the germ of p at 0.

Consider the map

$$\chi\colon \mathbb{R}^{m-1} \to A \to J^k(m, m-1), \quad y \mapsto j^k(\tilde{w}_y).$$

Its graph $\Gamma(\chi)$ is a subset of $\mathbb{R}^{m-1} \times J^k(m, m-1)$. The germ of Φ at a point $(y, j^k(\tilde{w})_y)$ equals the germ of $\Phi(x,z) = (\tilde{z}(x), z)$ at the point $(y, j^k(\tilde{w})_y)$, and as with p above, the germ of $(j^k\tilde{w})\tilde{}_y = \tilde{w}_y$ at y is $\mathscr{A}$-equivalent to the germ of $\tilde{w}$ at 0. We see that along the germ of $\Gamma(\chi)$ at $(0,w)$, the $\mathscr{A}$-type represented by Φ is constant, hence the space-germ is contained in $(*)$. Furthermore, the graph of a smooth function $\mathbb{R}^N \to \mathbb{R}^P$ will always be transverse to $\{0\} \times \mathbb{R}^P$ in $\mathbb{R}^N \times \mathbb{R}^P$. It follows that the set $(*)$ containing the graph of χ must be transverse to $\{0_{m-1}\} \times J^k(m, m-1)$, which proves the proposition. □

A $\mathscr{K}^k$-invariant submanifold Δ of $J^k(n,p)$ satisfies the *immersion condition* if for any stable map $F\colon X^{n+d} \to Y^{p+d}$, the restriction $F|\Delta_{\mathrm{source}}(F)$ is an immersion, where $\Delta_{\mathrm{source}}(F)$ is the subset $(j^kF)^{-1}\Delta$ consisting of points in source where the germ of F has k-jet in Δ.

Proposition 115. *The germ classes $E_{p,0}$ and $E_{p,0}(*)$ in $J^k(m, m-1)$ satsify the immersion condition, provided that k is sufficiently large.*

Proof. As we see in Proposition 112, $E_{p,0}$ and $E_{p,0}(*)$ are smooth submanifolds of jet space $J^k(m, m-1)$. If F is stable then j^kF is transverse to $\mathscr{K}^k$-orbits when k is large, and in particular to $E_{p,0}$ and $E_{p,0}(*)$, which pull back to submanifolds of source and thus, the statement makes sense.

Throughout the proof we use Δ to denote $E_{p,0}$ or $E_{p,0}(*)$, as the same argument will work in both cases.

We claim that it suffices to show that there exists one stable map

$$F\colon X^{2+d} \to Y^{1+d}$$

with $\Delta_{\text{source}}(F) \neq \emptyset$ such that the restriction $F|\colon \Delta_{\text{source}}(F) \to Y^{1+d}$ is an immersion:

Assume that F is such a map, that $x \in \Delta_{\text{source}}(F)$ and let $F(x) = y$. Let $\tilde{F}\colon \tilde{X}^{2+\tilde{d}} \to \tilde{Y}^{1+\tilde{d}}$ be another smooth, stable map with $\Delta_{\text{source}}(\tilde{F}) \neq 0$. Let $\tilde{x} \in \Delta_{\text{source}}(\tilde{F})$ and let $\tilde{y} = \tilde{F}(\tilde{x})$. The germ

$$\hat{F}_x\colon (X^{2+d}, x) \to (Y^{1+d}, y)$$

is stable and $\mathscr{E}\mathscr{K}$-equivalent to the stable germ

$$\tilde{F}_{\tilde{x}}\colon (\tilde{X}^{2+\tilde{d}}, \tilde{x}) \to (\tilde{Y}^{1+\tilde{d}}, \tilde{y}),$$

so for an appropriate choice of $d_1, d_2 \in \mathbb{N}_0$ (at least one of which can be taken as 0), the germs

$$\hat{F}_x \times \mathrm{id}_{(\mathbb{R}^{d_1},0)}\colon (X^{2+d}, x) \times (\mathbb{R}^{d_1}, 0) \to (Y^{1+d}, y) \times (\mathbb{R}^{d_1}, 0)$$

and

$$\tilde{F}_{\tilde{x}} \times \mathrm{id}_{(\mathbb{R}^{d_2},0)}\colon (\tilde{X}^{2+\tilde{d}}, \tilde{x}) \times (\mathbb{R}^{d_2}, 0) \to (\tilde{Y}^{1+\tilde{d}}, \tilde{y}) \times (\mathbb{R}^{d_2}, 0)$$

are $\mathscr{A}$-equivalent. Furthermore, $\Delta_{\text{source}}(\tilde{F} \times \mathrm{id}_{(\mathbb{R}^{d_2},0)})$ and $\Delta_{\text{source}}(F \times \mathrm{id}_{(\mathbb{R}^{d_1},0)})$ are diffeomorphic, and we see that $\hat{F}_x|(\Delta_{\text{source}}(F), x)$ is immersive if and only if $\tilde{F}_{\tilde{x}}|(\Delta_{\text{source}}(\tilde{F}), \tilde{x})$ is immersive.

Since being an immersion is a local property, this is enough.

But the ministable unfolding $F_{(p,0)}$ is an example of such a stable map F. Hence we are done. □

Now we look at the consequences of these results for the geometry of a stable map F.

Corollary 116. *Suppose that Δ is a germ class with the immersion condition – for example $E_{q,0}$ or $E_{q,0}(*)$. Given any stable map $F\colon N \to P$, the subsets $\Delta(F)$ are immersed submanifolds of $t(F)$.* □

Corollary 117. *Suppose that Δ is a germ class which contains at least one singularity with the immersion condition, such as $E_{q,0}$ or $E_{q,0}(*)$, and where all the contained germ classes define submanifolds of jet space. Given any stable map $F\colon N \to P$, the subsets $\Delta_{\text{strict}}(F)$ are embedded submanifolds of $t(F)$, and the restriction $F|\colon F^{-1}(\Delta_{\text{strict}}(F)) \cap \Sigma F \to \Delta_{\text{strict}}(F)$ restricts to a diffeomorphism on components.*

Proof. The subset $\Delta_{\text{strict}}(F)$ is an immersed submanifold of $t(F)$, by Corollary 116, and it cannot have any self intersections, as this would change the singularity type strictly presented. Hence it is embedded. □

One of the properties of the E-series singularities is that their positive instability locus is stratified by presentations of multigerms made up from $E_{q,0}$-singularities with $q \leq p$. We introduce a standard choice of coordinates for stable presentations of such multigerms (see also Lemma 11).

Given singularities $E_{p_1,0}(\lambda_1), \ldots, E_{p_r,0}(\lambda_r)$ we can construct a ministable multigerm $F_{(p_1,0).....(p_r,0)}$ such that the germ of $F_{(p_1,0).....(p_r,0)}$ at 0 in the target belongs to the $\mathscr{E}\mathscr{K}$-class $E_{p_1,0}(\lambda_1).\cdots.E_{p_r,0}(\lambda_r)$.

We know from (108) that the standard ministable unfolding of $E_{p_l,0}$ is a map $F_{(p_l,0)}\colon \mathbb{R}^{6p_l-1} \longrightarrow \mathbb{R}^{6p_l-2}$.

Define the map $F_{(p_1,0).\cdots.(p_r,0)}$ to be the disjoint union

$$(118)\qquad \bigsqcup_{i=1}^{r} F^i_{(p_1,0).\cdots.(p_r,0)}\colon \bigsqcup_{i=1}^{r} \mathbb{R}^{6p_i-1} \times \prod_{j\neq i, j=1,\ldots,r} \mathbb{R}^{6p_j-2} \longrightarrow \prod_{j=1,\ldots,r} \mathbb{R}^{6p_j-2}$$

where $F^i_{(p_1,0).\cdots.(p_r,0)}(a, b_1, \ldots, \hat{b}_i, \ldots, b_r) = (b_1, \ldots, F_{(p_i,0)}(a), \ldots, b_r)$.

Now $F_{(p_1,0).\cdots.(p_r,0)}$ is stable and its germ at $0 \in \prod_{j=1}^{k} \mathbb{R}^{6p_j-2}$ is $\mathscr{E}\mathscr{K}$-equivalent to $E_{p_1,0}(\lambda_1).\cdots.E_{p_r,0}(\lambda_r)$. Indeed, the map $F_{(p_1,0).\cdots.(p_r,0)}$ is a ministable unfolding for the singularity $E_{p_1,0}(\lambda_1).\cdots E_{p_r,0}(\lambda_r)$.

4.1.2. *Topology of the positive instability locus.* The *positive instability locus* of a weighted homogeneous map F, denoted $I(F^+)$, is the set of points $y \in t(F^+)$ where the germ of F^+ at $\Sigma F^+ \cap (F^+)^{-1}(y)$ is unstable. These are the points where a local smooth retraction $F_y\hat{} \longrightarrow (F^+)_y\hat{}$ cannot be found, and we have to resort to (E-)tame retractions to obtain our trivializations. For $E_{p,0}$ the positive instability locus has been parametrized by T. Wall and A. du Plessis through the following theorem.

Theorem 119. [dPW04, Theorem 3.2] *The instability locus of $F^k_{p,0}$ is the union of the images of the following deformations, for different choices of the $s \leq k$ and $c_i \geq 0$:*

$$(120)\quad y^3 + \lambda y \prod_{i=1}^{s} (x - \xi_i)^{2c_i} + \prod_{i=1}^{s} (x - \xi_i)^{3c_i}, \text{ where } \sum_{i=1}^{s} c_i = p, \text{ and } \sum_{i=1}^{s} c_i \xi_i = 0.$$

□

Here $F^k_{(p,0)}$ is the map obtained by removing the k lowest-weight unfolding parameters from $F_{(p,0)}$, and we note that $F^+_{(p,0)} = F^{p-1}_{(p,0)}$.

Unfortunately, the parametrization found in this theorem is not generally an embedding; in fact it is not even injective, and its differential at the origin is 0. However, we *can* show that its image is stratified smooth, and this is enough for our purposes.

Parametrization of the positive instability locus

Consider the instability locus of $F^+_{(p,0)}$, i.e. with $k = p - 1$.

We see that all of the deformations (120) are embedded in the deformation

$$D^+\colon (x,y,\xi_1,\dots,\xi_{p-2}) \mapsto \tag{121}$$
$$y^3 + \lambda y \prod_1^{p-2}(x-\xi_i)^2 \cdot \left(x + \tfrac{1}{2}\sum_{i=1}^{p-2}\xi_i\right)^4$$
$$+ \prod_1^{p-2}(x-\xi_i)^3 \cdot \left(x + \tfrac{1}{2}\sum_{i=1}^{p-2}\xi_i\right)^6 .$$

The different constellations (c_i) appearing in (120) correspond to subsets of type $\{\tilde{\xi}_i = \tilde{\xi}_j\}$ from the parameter space

$$\tilde{P} = \left\{ (\tilde{\xi}_1,\dots,\tilde{\xi}_{p-1}) | \tilde{\xi}_i = \xi_i \text{ for } i < p-1,\ \tilde{\xi}_{p-1} = -\frac{1}{p-2}\sum_{i=1}^{p-2}\xi_i \right\} \tag{122}$$

in the deformation (121). Each $(\xi_i) \in \mathbb{R}^{p-2}$ has a corresponding constellation $(c_i) \in \{0,1,\dots,p\}^s$ with $s \le p-1$.

We will see that the positive instability locus $I(F^+_{(p,0)})$ is stratified by submanifolds I_Δ, which are connected components of the presentation $\Delta_{\text{strict}}(F^+_{(p,0)})$ for $\Delta = \prod E_{c_i,0}(\lambda)$, where $\sum c_i = p$ and at least one of the c_i is ≥ 2. By Corollary 117, the presentations $\Delta_{\text{strict}}(F_{(p,0)})$ are embedded submanifolds of $t(F_{(p,0)})$; the $\Delta_{\text{strict}}(F_{(p,0)})$ define a stratification of $t(F_{(p,0)})$, and we note, furthermore, that the strata $\Delta_{\text{strict}}(F_{(p,0)})$ are ST-invariant.

When Theorem 119 states that the instability locus consists of the union of the images of the deformation given, it means that we can define a parametrization $p\colon \mathbb{R}^{p-2} \to t(F^+_{(p,0)})$ by setting

$$p(\xi_i) = (z(\xi_i), \underline{u}(\xi_i), \underline{v}(\xi_i)) .$$

Here the z, $\underline{u}$ and $\underline{v}$ are obtained through solving the equation

$$D^+(x,y,\xi_1,\dots,\xi_{p-2}) = \mathrm{pr}_1 \circ F^+(x,y,\underline{u},\underline{v})$$

by equating the coefficients of the two expressions viewed as polynomials in x and y, hence $z(\xi_k) =$ constant term, $u_i(\xi_k) = \mathrm{coef}(x^i)$ and $v_j(\xi_k) = \mathrm{coef}(yx^j)$.

Lemma 123. *Denote by $(c_i)_{i=1}^s$ the constellation of c_i associated to a fixed point $(\xi_1,\dots,\xi_{p-2})$ of the parameter space. At $p(\xi_i)$ we find presented an*

$$E_{c_1,0}(\lambda).\cdots.E_{c_s,0}(\lambda)$$

-singularity (modulo $\mathscr{E}\mathscr{K}$).

Proof. First we investigate the germ of $D^+_{(\xi_k)}$, given by

$$D^+_{(\xi_k)}(x,y) = D^+(x,y,(\xi_k))$$

at $(x,y) = (\tilde{\xi}_i, 0)$ for $i \in \{1,\dots,p-1\}$ by performing a change of variables $(x,y) \rightsquigarrow (\tilde{x},\tilde{y})$ where

$$\begin{cases} \tilde{x} = x + \tilde{\xi}_i, \\ \tilde{y} = y, \end{cases}$$

and considering the resulting germ at $(0,0)$. Here $\tilde{\xi}_i = \xi_i$ when $i < p-1$ and $\tilde{\xi}_{p-1} = -\frac{1}{p-2}\sum_{i=1}^{p-2}\xi_i$.

From the deformation (121) we see that $D^+_{(\xi_k)}(x,y)$ is just

$$\begin{aligned}
&\tilde{y}^3 + \lambda\tilde{y}(\textstyle\prod_{j\neq i}(\tilde{x} - \tilde{\xi}_i + \tilde{\xi}_j)^{2c_j})\tilde{x}^{2c_i} + (\prod_{j\neq i}(\tilde{x} - \tilde{\xi}_i + \tilde{\xi}_j)^{3c_j})\tilde{x}^{3c_i} \\
&= \tilde{y}^3 + \lambda\textstyle\prod_{j\neq i}(\tilde{\xi}_j - \tilde{\xi}_i)^{2c_j}\tilde{y}\tilde{x}^{2c_i} + \prod_{j\neq i}(\tilde{\xi}_j - \tilde{\xi}_i)^{3c_j}\tilde{x}^{3c_i} + \text{H.O.T.} \\
&\overset{\mathscr{A}-\text{equiv.}}{\rightsquigarrow} \tilde{y}^3 + \lambda(\textstyle\prod_{j\neq i}(\tilde{\xi}_j - \tilde{\xi}_i)^{2c_i})(\prod_{j\neq i}(\tilde{\xi}_j - \tilde{\xi}_i)^{3c_j})^{-\frac{2}{3}}\tilde{y}\tilde{x}^{2c_i} + \tilde{x}^{3c_i} + \text{H.O.T.} \\
&= \tilde{y}^3 + \lambda\tilde{y}\tilde{x}^{2c_i} + \tilde{x}^{3c_i} + \text{H.O.T.} \\
&\overset{\mathscr{A}-\text{equiv.}}{\rightsquigarrow} \tilde{y}^3 + \lambda\tilde{y}\tilde{x}^{2c_i} + \tilde{x}^{3c_i}.
\end{aligned}$$

where H.O.T. denotes the sum of terms of weight $> 3c_i$. We see that the germ of $D^+_{(\xi_i)}$ at $(x,y) = (\tilde{\xi}_i, 0)$ belongs to the class $E_{c_i,0}(\lambda)$.

Next, we link this result to the parametrization p. For any fixed point $(\xi_i) \in \mathbb{R}^{p-2}$ we get a fixed point $(\tilde{\xi}_1, \ldots, \tilde{\xi}_{p-1}) = (\xi_1, \ldots, \xi_{p-2}, -\frac{1}{2}\sum_{i=1}^{p-2}\xi_i) \in \mathbb{R}^{p-1}$ and the point $p(\xi_i) \in L(F^+_{(p,0)})$ denotes a point

$$(u,v) = (u_1, \ldots, u_{3p-2}, v_0, \ldots, v_{2p-1})(\xi_i)$$

in the parameter space U such that the map

$$F^1_{(p,0)}|_{(u,v)}(x,y) = \text{pr}_1 \circ F_{(p,0)}(x,y,u,v)$$

is exactly the same as the map $D^+_{(\xi_k)}(x,y)$, and in particular, the germ of $D^+_{(\xi_k)}(x,y)$ at $(\tilde{\xi}_i, 0)$ is the same as the germ of $F^1_{(p,0)}|_{(u,v)}(x,y)$ at $(\tilde{\xi}_i, 0)$, and this, again, is $\mathscr{E}\mathscr{K}$-equivalent to the germ of $F_{(p,0)}$ at $(\tilde{\xi}_i, 0, u, v)$, since $F_{(p,0)}$ is the standard stable unfolding.

But then the germ presented by $F_{(p,0)}$ at $p(\xi_i)$ is $\mathscr{E}\mathscr{K}$-equivalent to

$$E_{c_1,0}(\lambda).\cdots.E_{c_s,0}(\lambda),$$

and we are done. □

For each $\Delta = E_{c_1,0}(\lambda).\cdots.E_{c_s,0}(\lambda)$ we denote by P_Δ the subset

$$\{(\xi_1, \ldots, \xi_{p-2}) \in \mathbb{R}^{p-2} |\text{ the constellation } (c_i) \text{ is associated to } (\xi_1, \ldots, \xi_{p-2})\}.$$

Then we see that $p|P_\Delta$ parametrizes a set I_Δ in $t(F^+_{(p,0)})$ where we find Δ-singularities presented. We can also parametrize the corresponding subset $H_\Delta = (F^+_{(p,0)})^{-1}I_\Delta \cap \Sigma F^+_{(p,0)}$ of $s(F^+_{(p,0)})$. For any $E_{p,0}$ occurring in Δ, there is a map

$$\begin{gathered}
p_i \colon P_\Delta \to H_\Delta, \\
p_i(\xi_j) = (\tilde{\xi}_i, 0, u_1(\xi_j), \ldots, u_{3p-2}(\xi_j), v_0(\xi_j), \ldots, v_{2p-1}(\xi_j)), \\
\tilde{\xi}_i = \xi_i \text{ if } i < p-1,\ \tilde{\xi}_{p-1} = -\tfrac{1}{2}\textstyle\sum_{i=1}^{p-2}\xi_i,
\end{gathered}$$

parametrizing the component of H_Δ where the $E_{p,0}$-singularity is found.

Symmetries of the parametrization

Note that D^+, as a polynomial in the variables x and y as it is given in (121), is invariant under permutations of the ξ_i, $i = 1, \dots, p-2$. Thus, the same must hold for the parametrization p, and p is a symmetric map. It follows that the whole image of p is reached by the restriction of p to the subset $\{(\xi_1, \dots \xi_{p-2}) | \xi_1 \leq \dots \leq \xi_{p-2}\}$.

Note furthermore that S_{p-2}-invariance is not the only symmetry making a difference in (121). Consider subspaces of the parameter space $\tilde{P}$ as defined in (122) of the form $\{\tilde{\xi}_i = \tilde{\xi}_j\}$ where $i \neq j$ and $i, j < p-1$. In connection with these subspaces there will be another symmetry permuting $\tilde{\xi}_i = \tilde{\xi}_j$ and $\tilde{\xi}_{p-1}$. This symmetry only comes into play on some of the components of strata of dimension $< p-2$. Geometrically, this means that the instability locus will have self-intersections on some of the lower-dimensional strata.

The parametrization p cannot be an embedding since, due to its symmetries, it is not injective. It is not even immersive – for instance, in the case $p = 4$, immersivity fails along the stratum $P_{(2,0)^2}$, and for general p, immersivity certainly fails at $0 \in P$, as p does not have a linear part in either coordinate. However, we shall see that p restricts to a homeomorphism on components of P_Δ and that its image is stratified smooth with respect to the partition by strictly presented singularity types.

Stratification of the positive instability locus

By Corollary 117 the strict presentations by $F_{(p,0)}$ of $\prod E_{c_i,0}(*)$-singularities appearing in the instability locus are smooth. In fact, they can be parametrized in similar ways as the strata of the instability locus: The deformation

$$y^3 + y\underbrace{(w_0 + w_1 x + \dots + w_{p-2}x^{p-2})}_{W(x)}\prod_{i=1}^{p-1}(x-\xi_i)^{2c_i} + \prod_{i=1}^{p-1}(x-\xi_i)^{3c_i}$$

gives rise to a parametrization

$$\begin{aligned} \tilde{p}\colon \mathbb{R}^{p-2} \times \mathbb{R}^{p-1} &\to t(F_{(p,0)}), \\ (\xi_1, \dots \xi_{p-2}, w_0, \dots, w_{p-2}) &\mapsto (\underline{u}, \underline{v})(\xi_1, \dots \xi_{p-2}, w_0, \dots, w_{p-2}). \end{aligned}$$

where

$$\begin{aligned} u_i(\xi_1, \dots \xi_{p-2}, w_0, \dots, w_{p-2}) &= \operatorname{coef}(x^i), \\ v_j(\xi_1, \dots \xi_{p-2}, w_0, \dots, w_{p-2}) &= \operatorname{coef}(yx^j). \end{aligned}$$

We can imitate the proof of Lemma 123 to see that $\prod E_{c_i,0}$-singularities are presented by $F_{(p,0)}$ at the corresponding points $\tilde{p}(\xi_i, w_i)$ in the image, and by setting $w_0 = \lambda$ and $w_i = 0$ for $i > 0$ we see that the image of $\tilde{p}$ contains the positive instability locus of $F_{(p,0)}$. Following the proof by du Plessis and Wall [dPW04, Theorem 4.1] we see that in fact, the image of $\tilde{p}$ is transverse to the target of $t(F^+_{(p,0)})$ in $t(F_{(p,0)})$, which again we shall use to see that components of $\prod E_{c_i,0}(*)(F_{(p,0)})$ intersect $t(F^+_{(p,0)})$ in components of $I_{\prod E_{c_i,0}(\lambda)}$.

Lemma 124. *Let* $\prod E_{c_i,0}(\lambda)$ *be a singularity appearing in the positive stability locus* $I(F^+_{(p,0)})$. *Now, for* $\Delta = \prod E_{c_i,0}(*)$, *the components of* Y_Δ *passing through* $I(F^+_{(p,0)})$ *project submersively onto the non-positively weighted subspace of* $t(F_{(p,0)})$.

Remark 125. This result, along with the smoothness of the strata Y_Δ for $\Delta = \prod E_{c_i,0}(*)$ can already be found stated in the article by du Plessis and Wall [dPW04, Theorem 4.1]. However, there is a problem with the proof of smoothness of strata, which uses as a fact that the parametrization p is an embedding – and we have seen that it is not.

Proof. We see that the coefficients v_j for $2p \leq j \leq 3p-2$, which correspond to the non-positively weighted subspace of $t(F_{(p,0)})$, are linear in the variables w_j, and for any fixed $(\xi_i) \in \mathbb{R}^{p-2}$, the map $\mathbb{R}^{p-1} \to \mathbb{R}^{p-1}$ given by $(w_i)_{i=0}^{p-2} \mapsto (v_j((\xi_i),(w_i)))_{j=2p}^{3p-2}$ is linear with unitriangular matrix; hence it is a diffeomorphism. But then the map $\mathrm{pr}_{\mathbb{R}^{p-1}} \circ \tilde{p}$ is a submersion. □

Lemma 126. *The unstable source and target strata* H_Δ *and* I_Δ *are smooth, and have contractible components.*

Proof. Recall that we may assume that the components of P_Δ are subsets of

$$\{(\xi_1,\ldots,\xi_{p-2}) | \xi_1 \leq \xi_2 \leq \ldots \leq \xi_{p-2}\}$$

and that they are defined by relations of the type

$$(\xi_i = \xi_j) \text{ or } (\xi_k = -\frac{1}{2}\sum_{i=1}^{p-2} \xi_i).$$

Hence all relations of the type

(127)
$$(\xi_i \lesseqgtr \xi_j) \text{ or } (\xi_k \lesseqgtr -\frac{1}{2}\sum_{i=1}^{p-2} \xi_i)$$

will remain true throughout the component.

Note furthermore that H_Δ is a disjoint union

$$\bigsqcup p^i_{\text{source}}(\text{comp}(P_\Delta)),$$

where $\text{comp}(P_\Delta)$ goes through the components of P_Δ, and $i = 1,\ldots,p-1$. That is, $p^i_{\text{source}}(\text{comp}(P_\Delta))$ and $p^j_{\text{source}}(\text{comp}'(P_\Delta))$ will either coincide or be disjoint, as crossings will lead to strict presentation of other singularities than Δ.

We prove that $p^i_{\text{source}}|\text{comp}(P_\Delta)$ is injective, which will give smoothness and contractibility of components of H_Δ because the facts

a) $X_\Delta \cap \text{source}(F^+_{(p,0)})$ is smooth (to see this, note that $\mathrm{pr}_{\mathbb{R}^{p-1}}(X_\Delta) = \mathrm{pr}_{\mathbb{R}^{p-1}} \circ F_{(p,0)}(X_\Delta) = \mathrm{pr}_{\mathbb{R}^{p-1}}(Y_\Delta)$ and use Lemma 124), and
b) $\dim(X_\Delta \cap \text{source}(F^+_{(p,0)})) = \dim P_\Delta$

imply that $p^i_{\text{source}}|\text{comp}(P_\Delta)$ is an open topological embedding by invariance of domain.

It follows that the components of I_Δ are smooth and contractible by Corollary 117, since Δ has the immersion condition and H_Δ sits inside the source of the stable map $F_{(p,0)}$.

To prove that $p^i_{\text{source}}|\text{comp}(P_\Delta)$ is injective, we assume the opposite; then there exist

$$(128)\quad (\xi_1,\ldots,\xi_{p-2}) \neq (\bar{\xi}_1,\ldots,\bar{\xi}_{p-2}) \in \text{comp}(P_\Delta) \quad \text{s.t.} \quad p^i_{\text{source}}(\xi_j) = p^i_{\text{source}}(\bar{\xi}_j).$$

Note that by (128), in particular,

$$p(\xi_j) = p(\bar{\xi}_j) =: y$$

and

$$\begin{aligned} &\{p^i_{\text{source}}(\xi_j)|i = 1,\ldots,p-1\} \\ &= \{p^i_{\text{source}}(\bar{\xi}_j)|i = 1,\ldots,p-1\} \\ &= (F^+_{(p,0)})^{-1}(y) \cap \Sigma(F^+_{(p,0)}). \end{aligned}$$

Since $(\xi_j) \neq (\bar{\xi}_j)$, we must have $\xi_k \neq \bar{\xi}_k$ for some $k \in \{1,\ldots,p-2\}$, and the only possibility is

$$\begin{cases} \xi_k = -\frac{1}{2}\sum_{i=1}^{p-2}\bar{\xi}_i, \\ \bar{\xi}_k = -\frac{1}{2}\sum_{i=1}^{p-2}\xi_i. \end{cases}$$

We may assume that $\xi_k < \bar{\xi}_k$, which gives

$$\begin{cases} \xi_k < -\frac{1}{2}\sum_{i=1}^{p-2}\xi_i, \\ \bar{\xi}_k > -\frac{1}{2}\sum_{i=1}^{p-2}\bar{\xi}_i, \end{cases}$$

but this breaks the relations (127), hence (ξ_j) and $(\bar{\xi}_j)$ belong to different components of P_Δ, which gives a contradiction. It follows that $p^i_{\text{source}}|\text{comp}(P_\Delta)$ is injective, as wanted, and we are done. □

Corollary 129. *$F_{(p,0)}$ restricts to an injection on components of H_Δ, and hence the restriction $p|\colon \text{comp}(P_\Delta) \to I_\Delta$ is injective.*

Proof. To see that the injectivity of $F_{(p,0)}$ holds, just note that if the restriction $F_{(p,0)}|\text{comp}(H_\Delta)$ were to have double points, then the presented singularity would change, which is impossible inside the stratum I_Δ. The corollary must hold since $p^i_{\text{source}}|\text{comp}(P_\Delta)$ is injective for each i by the proof of the previous lemma, and $F_{(p,0)}$ restricts to an injection on components of H_Δ. □

Lemma 130. *Let $\Delta \in \{\prod_{i=1}^s E_{c_i,0}(\lambda)|\sum_{i=1}^s c_i = p, s < p\}$. Then the stratum $I_\Delta = I(F^+_{(p,0)}) \cap Y_\Delta$ is a union of components of $Y_\Delta \cap t(F^+_{(p,0)})$.*

Remark 131. Note that here, Y_Δ denotes the strict presentation of $\prod E_{c_i,0}(\lambda)$, not of $\prod E_{c_i,0}(*)$, and thus Y_Δ is a $\mathscr{K}$-class.

Proof. We can write

$$t(F^+_{(p,0)}) \cap Y_\Delta = T \cup U,$$

where T denotes the points of $t(F_{(p,0)})$ where the submanifolds $t(F^+_{(p,0)})$ and Y_Δ have nonempty, transverse intersection, and where U denotes the points of $t(F_{(p,0)})$ where the two submanifolds have non-transverse intersection.

Now by [Mat69a, Proposition 1.6], the subset U must coincide with the subset $I(F^+_{(p,0)}) \cap Y_\Delta$ of the positive instability locus. Since transverse intersection is an open property, T must be an open subset of $T \cup U$. Furthermore, the restriction

$$p|\colon \mathrm{comp}(P_\Delta) \to t(F^+_{(p,0)}) \cap Y_{\prod E_{c_i,0}(*)}$$

is an injective map by Corollary 129, and the dimensions of source and target are equal.

Hence $p|\mathrm{comp}(P_\Delta)$ is an open embedding by invariance of domain. Thus $U = I_\Delta = p(P_\Delta)$ has dimension $\dim P_\Delta = s-1$. Now, as a consequence of [Mat69a, Proposition 1.6], the singularity Δ cannot be presented stably and strictly by $F_{(p,0)}$ on a manifold of dimension other than $(6p-2) - \sum_{i=1}^{s}(6c_i - 2) = 2s-2$. We saw in Lemma 124 that the component of the presentation which intersects the instability locus must submerse onto the non-positively weighted subspace $\mathbb{R}^{p-1}$, so the presentation in $t(F^+_{(p,0)})$ must have dimension $(2s-2)-(p-1) = 2s-p-1$, which in particular is less than $s-1$, the dimension of U. Hence the only way that T can be open in $T \cup U$ is if T is a union of components of $T \cup U$. But then U is a union of components of $T \cup U$ as well; in fact it is the union of maximal-dimensional components of $t(F^+_{(p,0)}) \cap Y_\Delta$. □

We conclude:

Theorem 132. *The positive instability locus $I(F^+_{(p,0)})$ and its singular preimage $\Sigma F_{(p,0)} \cap F^{-1}_{(p,0)} I(F^+_{(p,0)})$ admit stratifications $\{I_\Delta\}$ and $\{H_\Delta\}$ into presentations, with smooth and contractible strata. These strata are components of the intersection of the presentations Y_Δ and X_Δ with the source and target of the positively weighted unfolding $F^+_{(p,0)}$.*

The parametrizations p^i_{source} and p restrict to homeomorphisms on components of strata, as does $F_{(p,0)}$. □

4.1.3. *E-tame retractions for $E_{1,0}$.* The stable unfolding for $E_{1,0}$ does not have any non-positively weighted unfolding variables, hence the retraction

$$(r_{(1,0)}, s_{(1,0)})\colon F_{(1,0)} \to F^+_{(1,0)}$$

is just the identities in source and target. We mention this because we *will* use these "retractions" in our later constructions.

4.1.4. *Construction of E-tame retractions for $E_{2,0}(*)$.* The weighted homogeneous ministable unfolding $F_{(2,0)}\colon \mathbb{R}^{11} \to \mathbb{R}^{10}$ of $E_{2,0}(\lambda)$ has one non-positively weighted unfolding variable v_4. The instability locus of $F^+_{(2,0)}\colon \mathbb{R}^{10} \to \mathbb{R}^9$ is just the origin by Theorem 119. Define the weighted homogeneous distance function $\rho\colon \mathbb{R}^{10} \to \mathbb{R}$ and use the level set restriction $(F_{(2,0)})_\epsilon\colon F^{-1}_{(2,0)}\rho^{-1}(\epsilon) \to \rho^{-1}(\epsilon)$ for some $\epsilon > 0$; now $(F^+_{(2,0)})_\epsilon$ is stable by [dPW95, Lemma 9.6.2]. Hence we can find a smooth retraction

$$(r_\epsilon, s_\epsilon)\colon (F_{(2,0)})_\epsilon \to (F^+_{(2,0)})_\epsilon,$$

and we can extend it to an E-tame retraction

$$(r_{(2,0)}, s_{(2,0)})\colon F_{(2,0)} \to F^+_{(2,0)}$$

using Lemma 36. Note, moreover, that the level set retractions were smooth; hence $(r_{(2,0)}, s_{(2,0)})$ will be stratified smooth – and we can find continuous, stratified smooth vector fields (ξ, η) in source and target that induce $(r_{(2,0)}, s_{(2,0)})$ as on p. 25.

The retraction we just constructed is $\mathbb{R}^+$-equivariant, but it does not necessarily have to be $\mathbb{R}^*$-equivariant since it is not necessarily $\{\pm 1\}$-equivariant. We can, however, do something about this:

The retraction $(r_{(2,0)}, s_{(2,0)})$, being smooth, is induced by the two vector fields ξ and η on $\mathbb{R}^{11}$ and $\mathbb{R}^{10}$, respectively, where η lifts $\frac{\partial}{\partial t}$ over $\mathrm{pr}_{\mathbb{R}}$ and ξ lifts η over $F_{(2,0)}$ as described on p. 25. We form two new vector fields $\tilde{\xi}$ and $\tilde{\eta}$ by setting

$$\tilde{\xi} = \tfrac{1}{2}(\xi + (-1)\cdot \xi),$$
$$\tilde{\eta} = \tfrac{1}{2}(\eta + (-1)\cdot \eta).$$

In fact, this average is just the Haar integral of ξ and η over the compact group $\{\pm 1\}$. Moreover, $\tilde{\xi}$ and $\tilde{\eta}$ are $\{\pm 1\}$-invariant, $\tilde{\eta}$ lifts $\frac{\partial}{\partial t}$ over pr because the $\mathbb{R}^*$-action on $\mathbb{R}$ is trivial (it has weight 0), and $\tilde{\xi}$ lifts η over F. The vector fields $\tilde{\xi}$ and $\tilde{\eta}$ are continuous and stratified smooth; hence they are integrable. Thus we obtain a new E-tame retraction $(\tilde{r}_{(2,0)}, \tilde{s}_{(2,0)})$ induced by $\tilde{\xi}$ and $\tilde{\eta}$. The oddly weighted components of $\tilde{\xi}$ and $\tilde{\eta}$ are zero, and thus the new retraction is identity on the oddly weighted coordinates, and $\mathbb{R}^*$-invariant on the evenly weighted ones; i.e. it is $\mathbb{R}^*$-equivariant.

We have proven:

Theorem 133. *There exists an $\mathbb{R}^*$-equivariant, stratified smooth, E-tame retraction $(r_{(2,0)}, s_{(2,0)})\colon F_{(2,0)} \to F^+_{(2,0)}$, and $F^+_{(2,0)}$ is topologically ministable.* □

Note that the retraction $(r_{(2,0)}, s_{(2,0)})$ and its foliation are not uniquely defined, as we have made a choice as to which smooth retraction to use on the level set.

4.1.5. *Construction of E-tame retractions for $F_{3,0}(*)$.* The weighted homogeneous ministable unfolding $F_{(3,0)}\colon \mathbb{R}^{17} \to \mathbb{R}^{16}$ of $E_{3,0}(\lambda)$ has two non-positively weighted unfolding variables, v_7 and v_6. By Theorem 119 the instability locus of $F^+_{(3,0)} = F^2_{(3,0)}\colon \mathbb{R}^{15} \to \mathbb{R}^{14}$ is 1-dimensional and has two types of singularities presented in it; namely an $E_{3,0}(\lambda)$ singularity at the origin and $E_{2,0}(\lambda).E_{1,0}$ singularities along a 1-dimensional, $\mathbb{R}^+$-invariant submanifold of $\mathbb{R}^{14}$.

Let ρ be the weighted distance function in $t(F_{(3,0)})$, let $\epsilon > 0$ and denote by $(F_{(3,0)})_\epsilon$ the restriction to the level sets $(F^{-1}_{(3,0)}\rho^{-1}(\epsilon), \rho^{-1}(\epsilon))$. By [dPW95, Lemma 9.6.2], $(F^1_{(3,0)})_\epsilon$ is stable and the instability locus of $(F^2_{(3,0)})_\epsilon$ is $\rho^{-1}(\epsilon) \cap I(F^2_{(3,0)})$, which consists of two points y_1, y_2 where an $E_{2,0}(\lambda).E_{1,0}$-singularity is presented, and the germ of $(F^1_{(3,0)})_\epsilon$ at y_i is $\mathscr{A}$-equivalent to the multigerm $F_{(2,0).(1,0)}\colon \mathbb{R}^{15} \to \mathbb{R}^{14}$.

Choose weighted homogeneous coordinates (ψ_{y_i}, ϕ_{y_i}) at y_i such that:

$$\begin{array}{ccc} \left(s((F^1_{(3,0)})_\epsilon), F^{-1}_{(3,0)}(y_i) \cap \Sigma F_{(3,0)}\right) & \xrightarrow{(F^1_{(3,0)})_\epsilon} & (t((F^1_{(3,0)})_\epsilon), y_i) \\ \downarrow \psi_{y_i} & & \downarrow \phi_{y_i} \\ \left(s(F_{(2,0).(1,0)}), \bigsqcup_2 0\right) & \xrightarrow[F_{(2,0).(1,0)}]{} & (t(F_{(2,0).(1,0)}), 0) \end{array}$$

By Lemma 29 and the previous case there exists an E-tame retraction

$$(r_{(2,0).(1,0)}, s_{(2,0).(1,0)})\colon F_{(2,0).(1,0)} \to F^+_{(2,0).(1,0)}.$$

The fiber $\phi^{-1}_{y_i}(s^{-1}_{(2,0).(1,0)}(0))$ coincides with the presentation of $E_{2,0}(*).E_{1,0}$; hence the fiber of the total induced retraction $t((F_{(3,0)})_\epsilon) \to \phi^{-1}_{y_i}(t(F^+_{(2,0).(1,0)}))$ is transverse to $t((F^+_{(3,0)})_\epsilon)$ by Lemma 124. By Proposition 26 this retraction induces an E-tame retraction $(F_{(3,0)})_\epsilon \to (F^+_{(3,0)})_\epsilon$ near y_i, which has the same fibers.

On the other hand, away from y_i we can find smooth retractions $(F_{(3,0)})\hat{_\epsilon} \to (F^+_{(3,0)})\hat{_\epsilon}$. Using Lemma 32 we may combine the two retractions to obtain an E-tame retraction $(F_{(3,0)})_\epsilon \to (F^+_{(3,0)})_\epsilon$ and by Lemma 36 we obtain an E-tame retraction

$$(r_{(3,0)}, s_{(3,0)})\colon F_{(3,0)} \to F^+_{(3,0)}.$$

Note again that, in spite of our notation, our method does not define the retraction $(r_{(3,0)}, s_{(3,0)})$ uniquely, as we have made choices regarding which smooth retractions to use in the very beginning.

We have proven:

Theorem 134. *There exists an E-tame retraction*

$$(r_{(3,0)}, s_{(3,0)})\colon F_{(3,0)} \to F^+_{(3,0)},$$

and $F^+_{(3,0)}$ is topologically ministable. □

4.1.6. *Construction of E-tame retractions for $E_{4,0}(*)$.* The weighted homogeneous ministable unfolding $F_{(4,0)}\colon \mathbb{R}^{23} \to \mathbb{R}^{22}$ has three non-positively weighted unfolding variables v_8, v_9, v_{10}, and the instability locus is the stratified set which we analyzed in Chapter 4.1.2. We will construct the retraction $F_{(4,0)} \to F^+_{(4,0)}$ by first restricting to a level set of the weighted homogeneous distance function ρ, on which we find presented combinations of E-singularities, all with $p < 4$, suggesting an inductive construction of the retraction. We can find local E-tame retractions using the previous results, but in order to combine them we need to control the geometry near tho instability locus.

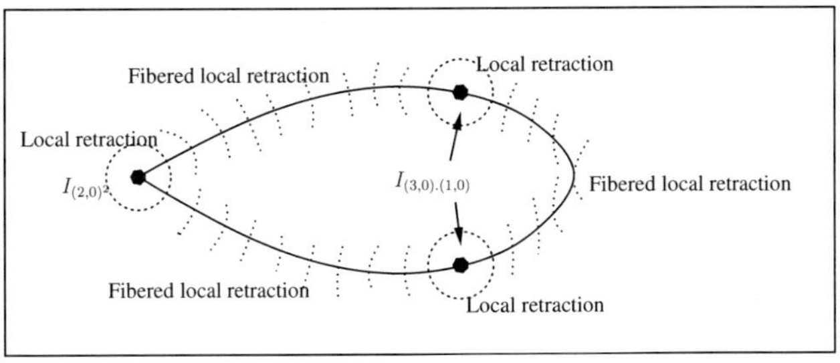

FIGURE 5. *Target situation for* $E_{4,0}(*)$. We choose a tubular neighborhood about the 1-dimensional stratum, such that its fibers coincide with level sets of the weighted distance function about the 0-dimensional strata. We can find local retractions near the 0-dimensional strata and in the tubular fibers, and we can create a global retraction by forcing the local retractions to coincide.

We find a tubular neighborhood about the 1-dimensional stratum, whose fibers extend the fibration near the 0-dimensional strata by level sets of local weighted homogeneous distance functions. We can find local E-tame retractions in the fibers of the tubular neighborhood as well as at the zero-dimensional strata. Rather than glue the local retractions together, we can try to force them to coincide.

We choose coordinates in the fibers of the tubular neighborhood about the stratum by using the contractibility of $\mathscr{A}_f/MC(\mathscr{A}_f)$ for the stable multigerm $f = F_{(2,0).(1,0)^2}$. In this way we ensure that the local retractions defined on the fibers coincide with the local retractions defined near the 0-dimensional strata, producing an E-tame retraction in a neighborhood of the instability locus.

Finally, we combine this retraction with any smooth retraction off the instability locus.

Let us look at the details:

Pass to a slice

Let $\rho_{(4,0)}\colon \mathbb{R}^{22} \to \mathbb{R}$ be the weighted distance function as defined in Chapter 2.4, pick $\epsilon > 0$, and restrict to the level sets $(F^{-1}_{(4,0)}\rho^{-1}_{(4,0)}(\epsilon), \rho^{-1}_{(4,0)}(\epsilon))$. We denote the restricted map by $(F_{(4,0)})_\epsilon$; similarly we denote by $(F^+_{(4,0)})_\epsilon$ the restriction of $F^+_{(4,0)}$ to

$$\left(F^{-1}_{(4,0)}\rho^{-1}_{(4,0)}(\epsilon) \cap s(F^+_{(4,0)}), \rho^{-1}_{(4,0)}(\epsilon) \cap t(F^+_{(4,0)})\right).$$

If we can find an E-tame retraction $(r, s)\colon (F_{(4,0)})_\epsilon \to (F^+_{(4,0)})_\epsilon$, then we can find an E-tame retraction $(R, S)\colon F_{(4,0)} \to F^+_{(4,0)}$ by Lemma 36.

By [dPW95, Lemma 9.6.2], $(F^1_{(4,0)})_\epsilon$ is stable, and we can find a smooth retraction $(F_{(4,0)})_\epsilon \to (F^1_{(4,0)})_\epsilon$. By [dPW95, Lemma 9.3.22], it suffices to find an E-tame

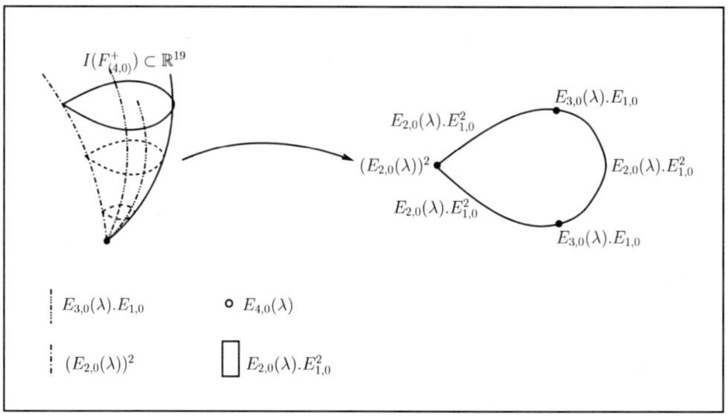

FIGURE 6. *Target situation.* The stratified instability locus $I(F^+_{(4,0)})$ before and after intersection with the level set $\rho^{-1}_{(4,0)}(\epsilon)$.

retraction $(F^1_{(4,0)})_\epsilon \to (F^+_{(4,0)})_\epsilon$ in order to get an E-tame retraction $(F_{(4,0)})_\epsilon \to (F^+_{(4,0)})_\epsilon$.

The instability locus $I(F^+_{(4,0)})$ of $F^+_{(4,0)}$ is $\mathbb{R}^+$-invariant, hence it is cut transversely by $\rho^{-1}_{(4,0)}(\epsilon)$. The instability locus $I((F^+_{(4,0)})_\epsilon)$ equals $I(F^+_{(4,0)}) \cap \rho^{-1}(\epsilon)$ [dPW95, Lemma 9.6.2], hence it is a 1-dimensional subset of $\rho^{-1}_{(4,0)}(\epsilon)$, and it is stratified by smooth submanifolds $I_{(2,0)^2}$, $I_{(2,0).(1,0)^2}$ and $I_{(3,0).(1,0)}$ of $\rho^{-1}_{(4,0)}(\epsilon)$, which are components of the strict presentations of

$$(E_{2,0}(\lambda))^2, \quad E_{2,0}(\lambda).E^2_{1,0} \quad \text{and} \quad E_{3,0}(\lambda).E_{1,0}$$

in $t((F^+_{(4,0)})_\epsilon)$, of dimension 0, 1 and 0, respectively, as seen in Figure 6.

For simplicity, we agree to denote $(F^1_{(4,0)})_\epsilon$ by F, and $(F^+_{(4,0)})_\epsilon$ by F^+ through the whole construction of E-tame retractions for $E_{4,0}(*)$.

Finding local tubular neighborhoods of $(H_{(2,0).(1,0)^2}, I_{(2,0).(1,0)^2})$ near the 0-dimensional strata $(H_{(2,0)^2}, I_{(2,0)^2})$ and $(H_{(3,0).(1,0)}, I_{(3,0).(1,0)})$

Recall the notation $H_\Delta = (F^+)^{-1}(I_\Delta) \cap \Sigma F^+$.

Let $\tilde{y} \in I_\Delta$, where $\Delta \in \{(3,0).(1,0), (2,0)^2\}$. The germ of F at $\tilde{y}$ is $\mathscr{A}$-equivalent to the germ F_Δ, which is weighted homogeneous. Here F_Δ is defined as in (118), that is:

$$\begin{aligned} F_{(2,0)^2} = \quad & F_{(2,0)} \times \mathrm{id}_{t(F_{(2,0)})} \bigsqcup \mathrm{id}_{t(F_{(2,0)})} \times F_{(2,0)} : \\ & s(F_{(2,0)}) \times t(F_{(2,0)}) \bigsqcup t(F_{(2,0)}) \times s(F_{(2,0)}) \to t(F_{(2,0)}) \times t(F_{(2,0)}) \\ F_{(3,0).(1,0)} = \quad & F_{(3,0)} \times \mathrm{id}_{t(F_{(1,0)})} \bigsqcup \mathrm{id}_{t(F_{(3,0)})} \times F_{(1,0)} : \\ & s(F_{(3,0)}) \times t(F_{(1,0)}) \bigsqcup t(F_{(3,0)}) \times s(F_{(1,0)}) \to t(F_{(3,0)}) \times t(F_{(1,0)}) \end{aligned}$$

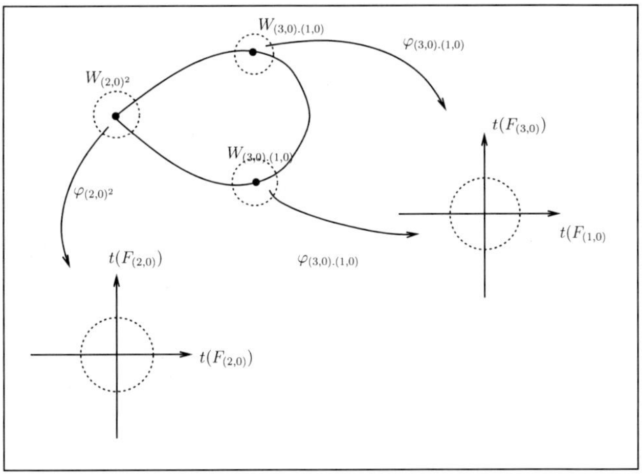

FIGURE 7. *Target situation.* Choice of target coordinates near $I_{(2,0)^2}$ and $I_{(3,0).(1,0)}$.

and, in particular,

$$\begin{aligned} s(F_{(2,0)^2}) &= s(F_{(2,0)}) \times t(F_{(2,0)}) \bigsqcup t(F_{(2,0)}) \times s(F_{(2,0)}) \\ t(F_{(2,0)^2}) &= t(F_{(2,0)}) \times t(F_{(2,0)}) \\ s(F_{(3,0).(1,0)}) &= s(F_{(3,0)}) \times t(F_{(1,0)}) \bigsqcup t(F_{(3,0)}) \times s(F_{(1,0)}) \\ t(F_{(3,0).(1,0)}) &= t(F_{(3,0)}) \times t(F_{(1,0)}). \end{aligned}$$

Fix local coordinates

$$\begin{aligned} &\varphi_\Delta \colon W_\Delta \to U_\Delta \subset t(F_\Delta), \text{ and} \\ &\psi_\Delta \colon V_\Delta \to \tilde{U}_\Delta \subset s(F_\Delta), \end{aligned}$$

such that W_Δ is a neighborhood of $\tilde{y}$ and $\varphi_\Delta(\tilde{y}) = 0$, and V_Δ is a neighborhood of $F^{-1}(\tilde{y}) \cap \Sigma F$, with $\psi_\Delta(F^{-1}(\tilde{y}) \cap \Sigma F) = \bigsqcup_2 \{0\}$, and such that $\varphi_\Delta \circ F = F_\Delta \circ \psi_\Delta$. See Figure 7.

Define a distance function in $t(F_\Delta)$ as follows. Denote $\Delta = (p_1, 0).(p_2, 0) \in \{(2,0)^2, (3,0).(1,0)\}$. Define $\rho_\Delta \colon t(F_\Delta) \to \mathbb{R}$, $\rho_\Delta = \rho_{(p_1,0)} \circ \mathrm{pr}_{t(F_{(p_1,0)})}$. Now

$$\rho_\Delta^{-1}(\epsilon) = \rho_{(p_1,0)}^{-1}(\epsilon) \times t(F_{(p_2,0)}) \subset t(F_{(p_1,0)}) \times t(F_{(p_2,0)}) = t(F_{(p_1,0).(p_2,0)}).$$

These level sets are transverse in $t(F_\Delta)$ to any $\mathbb{R}^+$-invariant manifold not contained in $\{0\} \times t(F_{(p_2,0)})$, so in particular to $\varphi_\Delta(I_{(2,0).(1,0)^2})$. (We arrange our coordinates so that $\varphi_\Delta(I_{(2,0).(1,0)^2})$ is contained in $t(F_{(p_1,0)}) \times \{0\}$.)

Taking these level sets as fibers and passing back to W_Δ using φ_Δ^{-1}, we define a tubular neighborhood (T_Δ, π_Δ) of $I_{(2,0).(1,0)^2} \cap W_\Delta$ in W_Δ. Using the preimages of the fibers under F, we get a corresponding tubular neighborhood $(\tilde{T}_\Delta, \tilde{\pi}_\Delta)$ of $H_{(2,0).(1,0)^2}$

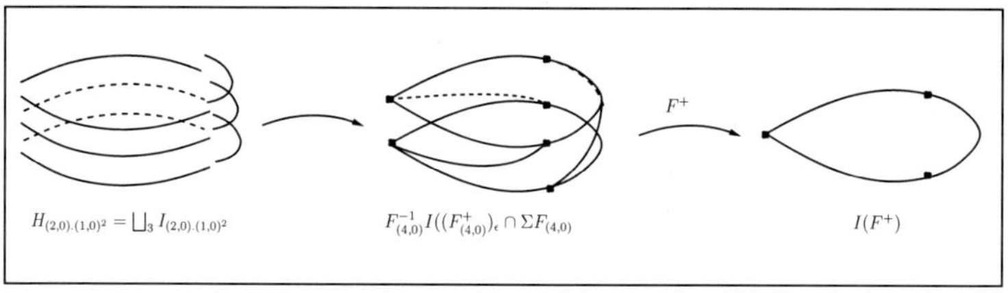

FIGURE 8. The source stratum $H_{(2,0).(1,0)^2}$ is diffeomorphic to $\sqcup_3 I_{(2,0).(1,0)^2}$.

in V_Δ:

$$\begin{array}{ccccccc}
\tilde{T}_\Delta \subset V_\Delta & & & \xrightarrow{\quad F| \quad} & & & T_\Delta \subset W_\Delta \\
& \searrow^{\psi_\Delta} & & & & \swarrow^{\varphi_\Delta} & \\
& & s(F_\Delta) & \xrightarrow[F_\Delta]{} & t(F_\Delta) & & \\
\downarrow \tilde{\pi}_\Delta & & \downarrow \rho_\Delta \circ F_\Delta & & \downarrow \rho_\Delta & & \downarrow \pi_\Delta \\
& & [0,\epsilon] & \xrightarrow[\mathrm{id}]{} & [0,\epsilon] & & \\
& \swarrow & & & & \searrow & \\
H_{(2,0).(1,0)^2} \cong \bigsqcup_3 I_{(2,0).(1,0)^2} & & & \xrightarrow[\bigsqcup_3 \mathrm{id}]{} & & & I_{(2,0).(1,0)^2}
\end{array}$$

That $(\tilde{T}_\Delta, \tilde{\pi}_\Delta)$ is a tubular neighborhood follows from the following lemma:

Lemma 135. *Any tubular neighborhood of (an open subset A of) $I_{(2,0).(1,0)^2}$ pulls back to a tubular neighborhood of ($F^{-1}A \subset$) $H_{(2,0).(1,0)^2}$ over F, and they are both trivial as bundles. The map F is trivial over the restriction to any fiber.*

Proof. Assume that A is an open subset of $I_{(2,0).(1,0)}$, possibly all of it. Let (T, π) be a tubular neighborhood of A. By Corollary 117, $H_{(2,0).(1,0)^2}$ is diffeomorphic to $\bigsqcup_3 I_{(2,0).(1,0)^2}$ and F restricts to a diffeomorphism F^i on each component $H^i_{(2,0).(1,0)^2}$ of $H_{(2,0).(1,0)^2}$. See Figure 8. On one of these components, F restricts to a germ of singularity type $E_{2,0}(\lambda)$, while on the other two its singularity type is $E_{1,0}$.

The map $\tilde{\pi} \colon \tilde{T} := F^{-1}T \to F^{-1}A$ defined by $\tilde{\pi} = \bigsqcup_{i=1}^{3} (F^i)^{-1} \circ \pi \circ F$ is a retraction, hence a submersion, near A. Hence, for a sufficiently small neighborhood W of A

in $\tilde{T}$, the fiber $\tilde{\pi}^{-1}(x) \cap W$ is a smooth submanifold of W. Furthermore, the fibers of $\tilde{\pi}$ are the preimages over F of the fibers of π.

We may assume (by shrinking W) that each component of W contains one component of A, and since all components of $H_{(2,0).(1,0)^2}$ and $I_{(2,0).(1,0)^2}$ (and also of $F^{-1}A$ and A since the $H_{(2,0).(1,0)^2}$ and $I_{(2,0).(1,0)^2}$ are 1-dimensional) are contractible by Lemma 126, we know that $(W, \tilde{\pi})$ and (T, π) are trivial bundles. The restriction $F|\colon W \to T$ is an unfolding of each restriction $F|\colon \tilde{\pi}^{-1}F^{-1}(y) \cap W \to \pi^{-1}(y)$, each of which is stable [dP, Proposition 2.4]. But then, by [GWdPL76, Lemma 3.2], $F|\colon W \to T$ is trivial over $F|\colon \tilde{\pi}^{-1}F^{-1}(y) \cap W \to \pi^{-1}(y)$. Since $F|\colon (\tilde{\pi})^{-1}(x) \cap W \to \pi^{-1}(F(x))$ is submersive outside $\bigsqcup_3\{0\}$ for each $x \in A$, we may assume (by shrinking T) that $W = \tilde{T}$. □

Schematically, this means that we can decompose $F|$ over $I_{(2,0).(1,0)^2} \cap W_\Delta$ as follows (setting $\tilde{\Delta} = (2,0).(1,0)^2$):

$$\begin{array}{ccc} V_\Delta & \xrightarrow{F|} & W_\Delta \\ \downarrow \cong & & \downarrow \cong \\ \bigsqcup_3 \tilde{L} \times (I_{\tilde{\Delta}} \cap W_\Delta) & \xrightarrow[F_L \times \mathrm{id}|I_{\tilde{\Delta}} \cap W_\Delta]{} & L \times (I_{\tilde{\Delta}} \cap W_\Delta) \end{array} \tag{136}$$

where $F_L = F|\colon \bigsqcup_3 \tilde{L} \cong (F|)^{-1}L \to L$, and where L is the fiber of (T, π).

Find global tubular neighborhoods of $(H_{(2,0).(1,0)^2}, I_{(2,0).(1,0)^2})$

Again, assume $\tilde{y} \in I_\Delta$, with $\Delta \in \{(2,0)^2, (3,0).(1,0)\}$.

We pick open neighborhoods $\tilde{W}_\Delta$ and $\tilde{V}_\Delta$ of I_Δ and H_Δ, respectively, such that $\mathrm{cl}(\tilde{W}_\Delta) \subset W_\Delta$ and $\mathrm{cl}(\tilde{V}_\Delta) \subset V_\Delta$. We are going to construct global tubular neighborhoods of $(H_{(2,0).(1,0)^2}, I_{(2,0).(1,0)^2})$ which coincide with the local tubular neighborhoods $\big((\tilde{T}_\Delta, \tilde{\pi}_\Delta), (T_\Delta, \pi_\Delta)\big)$ in $(\tilde{W}_\Delta, \tilde{V}_\Delta)$.

Proposition 137. *We can find a trivial tubular neighborhood (T, π) of the stratum $I_{(2,0).(1,0)^2}$ in $t(F)$ which pulls back to a tubular neighborhood $F^{-1}T =: \tilde{T}$ of $H_{(2,0).(1,0)^2}$ in $s(F)$ with retraction $\tilde{\pi}$, such that the fibers of $(\tilde{\pi}, \pi)$ coincide with the fibers of $(\tilde{\pi}_\Delta, \pi_\Delta)$ near (H_Δ, I_Δ), where $\Delta \in \{(2,0)^2, (3,0).(1,0)\}$.*

Furthermore, the restriction of F to any fiber is stable, and F is trivial over $I_{(2,0).(1,0)^2}$.

Proof. We note that by finding suitable local sprays and combining them using a partition of unity (see p. 20), we can find a trivial tubular neighborhood $e\colon L \times I_{(2,0).(1,0)^2} \to T$ of $I_{(2,0).(1,0)^2}$ in $t(F)$ such that near I_Δ, the fibers $e(L \times \{z\})$ of the tubular neighborhood coincide with $\varphi_\Delta^{-1}(\rho_\Delta^{-1}(\epsilon))$. See Figure 9. By Lemma 135, this

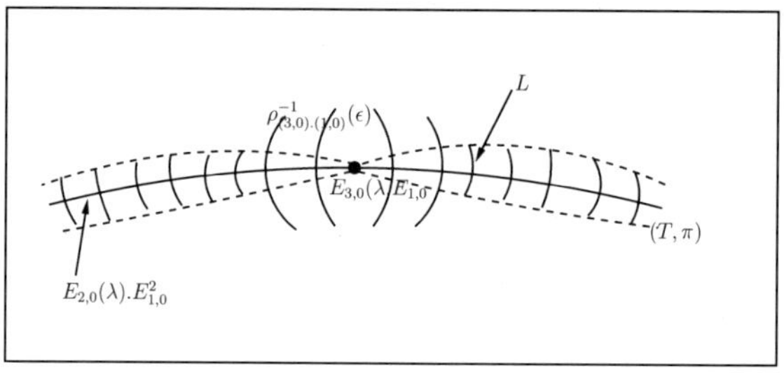

FIGURE 9. Inside $\rho^{-1}(\epsilon)$, we form a tubular neighborhood about the stratum $(E_{2,0}(\lambda).E^2_{1,0})_{\text{strict}}(F^+)$, whose fibers near $I_{(3,0).(1,0)}$ coincide with level sets of $\rho_{(3,0).(1,0)}$.

tubular neighborhood pulls back to a trivial tubular neighborhood $(\tilde{T}, \tilde{\pi})$, with the required properties near the H_Δ.

The restrictions of F to fibers $\tilde{\pi}^{-1}F^{-1}(y)$ are stable by [dP, Proposition 2.4], because they are transverse to $I_{(2,0).(1,0)^2}$, which is just the presentation of $E_{2,0}(\lambda).E^2_{1,0}$ by F. Let us denote the fiber map by F_L for now; it is $\mathscr{E}\mathscr{K}$-equivalent to $F_{(2,0).(1,0)^2}$.

As in (136), we can decompose F_T over $I_{(2,0).(1,0)^2}$ as follows:

$$\begin{array}{ccc} F^{-1}_{(4,0)}(T) = \tilde{T} & \xrightarrow{F=:F_T} & T \\ \downarrow \cong & & \downarrow \cong \\ \bigsqcup_3 \tilde{L} \times I_{\tilde{\Delta}} & \xrightarrow[F_L \times \mathrm{id}|_{I_{\tilde{\Delta}}}]{} & L \times I_{\tilde{\Delta}} \end{array}$$

setting $\tilde{\Delta} = (2,0).(1,0)^2$, where $F_L = \bigsqcup_{i=1}^3 F_L^i \colon \bigsqcup_3 \tilde{L} \to L$. □

By only restricting to $(\tilde{T}, T)$ and trivializing F_T over the source and target strata $(H_{(2,0).(1,0)^2}, I_{(2,0).(1,0)^2})$, we are allowing one dimension too much to be able to really control the geometry:

Since the dimension of L is $22 - 1 - 1 - 1 = 19$ and the dimension of $t(F_{(2,0).(1,0)^2})$ is $10 + 4 + 4 = 18$, we see that F_L is $\mathscr{A}$-equivalent to

$$F_{(2,0).(1,0)^2} \times \mathrm{id}_{\mathbb{R}},$$

giving a fixed set of coordinates (ψ_L, ϕ_L) such that

$$\phi_L(L) = t(F_{(2,0)}) \times t(F_{(1,0)}) \times t(F_{(1,0)}) \times \mathbb{R} \tag{138}$$

and

$$\psi_L\Big(\bigsqcup_3 \tilde{L}\Big) = \left(s(F_{(2,0)}) \times t(F_{(1,0)^2}) \sqcup \bigsqcup_2 t(F_{(2,0)}) \times s(F_{(1,0)}) \times t(F_{(1,0)}) \right) \times \mathbb{R},$$

such that

$$\begin{array}{ccc} \bigsqcup_3 \tilde{L} & \xrightarrow{\quad F_L \quad} & L \\ \Big\downarrow{\scriptstyle \psi_L} & & \Big\downarrow{\scriptstyle \phi_L} \\ s(F_{(2,0).(1,0)^2}) \times \mathbb{R} & \xrightarrow[F_{(2,0).(1,0)^2} \times \mathrm{id}_{\mathbb{R}}]{} & t(F_{(2,0).(1,0)^2}) \times \mathbb{R} \end{array}$$

The extra $\mathbb{R}$-component causes F_L not to be ministable as needed in order to apply the results of Chapter 3 for interpolating between coordinate charts in the fibers. We solve the problem by factoring the $\mathbb{R}$-component out by restricting to codimension 1 submanifolds $(\tilde{M}, M)$ chosen such that the restriction of F is still stable. Close to the 0-dimensional strata I_Δ we want M to coincide with $t(F^1_\Delta)$, and similarly in source, as described in the statement of the next proposition.

Proposition 139. *We can find a submanifold M of T which pulls back to a submanifold $F^{-1}M =: \tilde{M}$ of $\tilde{T}$, both of codimension 1, such that*

$$(140)\quad \begin{array}{lcl} \varphi_{(2,0)^2}(M \cap \tilde{W}_{(2,0)^2}) & = & t(F_{(2,0)}) \times t(F^+_{(2,0)}) \cap \varphi_{(2,0)^2}(T), \\ \varphi_{(3,0).(1,0)}(M \cap \tilde{W}_{(3,0).(1,0)}) & = & t(F_{(1,0)}) \times t(F^1_{(3,0)}) \cap \varphi_{(3,0).(1,0)}(T), \\ \psi_{(2,0)^2}(\tilde{M} \cap \tilde{V}_{(2,0)^2}) & = & \Big(s(F_{(2,0)}) \times t(F^+_{(2,0)}) \sqcup t(F_{(2,0)}) \times s(F^+_{(2,0)})\Big) \\ & & \cap \psi_{(2,0)^2}(\tilde{T}), \\ \psi_{(3,0).(1,0)}(\tilde{M} \cap \tilde{V}_{(3,0).(1,0)}) & = & \Big(s(F_{(1,0)}) \times t(F^1_{(3,0)}) \sqcup t(F_{(1,0)}) \times s(F^1_{(3,0)})\Big) \\ & & \cap \psi_{(3,0).(1,0)}(\tilde{T}), \end{array}$$

and such that the restriction $F_M := F_{(4,0)}|\colon \tilde{M} \to M$ is stable.

Remark 141. We agree to write $F^1_{(2,0)^2}$ for $F_{(2,0)} \times \mathrm{id}_{t(F^+_{(2,0)})} \bigsqcup \mathrm{id}_{t(F_{(2,0)})} \times F^+_{(2,0)}$.

Proof. Note that $\phi_L^{-1}(\{0\} \times \mathbb{R}) \subset L$ as defined in (138) equals the presentation $E_{2,0}(\lambda).E^2_{1,0}(F_L)$. Denote by

$$\tilde{e}\colon (s(F_{(2,0).(1,0)^2}) \times \mathbb{R} \times I_{(2,0).(1,0)^2}, t(F_{(2,0).(1,0)^2}) \times \mathbb{R} \times I_{(2,0).(1,0)^2}) \to (\tilde{T}, T)$$

the tubular neighborhood embeddings. We denote the image $\tilde{e}(\{0\} \times \mathbb{R} \times I_{(2,0).(1,0)^2})$ by $T_{(2,0).(1,0)^2}$. Now

$$M_{\tilde{\Delta}} = \tilde{e}(t(F_{(2,0).(1,0)^2}) \times \{0\} \times I_{(2,0).(1,0)^2})$$

is a codimension 1 submanifold of T. Still writing

$$\Delta = (p_1, 0).(p_2, 0) \in \{(2,0)^2, (3,0).(1,0)\},$$

we let M_{Δ} denote the codimension 1 submanifold of $T \cap W_{\Delta}$ defined by

$$\varphi_{\Delta}^{-1}(t(F_{(p_1,0)}) \times t(F^1_{(p_2,0)})).$$

Note that both $M_{\tilde{\Delta}} \cap L_y$ and $M_{\Delta} \cap L_y$ are transverse to $T_{(2,0).(1,0)^2} \cap L_y$.

Form smooth retractions $s_{\tilde{\Delta}}, s_{\Delta} \colon T \cap W_{\Delta} \to T_{(2,0).(1,0)^2} \cap W_{\Delta}$ such that

$$s_i^{-1}(I_{(2,0).(1,0)^2}) = M_i, \quad i = \tilde{\Delta}, \Delta.$$

Using the non-relative version of Lemma 32 as described in Remark 33, we find a new retraction $S_{\Delta} \colon T \cap W_{\Delta} \to T_{(2,0).(1,0)^2} \cap W_{\Delta}$ which agrees with s_{Δ} in $\tilde{W}_{\Delta}$, and which agrees with $s_{\tilde{\Delta}}$ away from I_{Δ}; obviously this is the restriction of a global retraction

$$S \colon T \to T_{(2,0).(1,0)^2}.$$

The preimage $M := S^{-1}(I_{(2,0).(1,0)^2})$ is a smooth, codimension 1 submanifold of T satisfying the conditions (140) of the proposition near I_{Δ}.

M pulls back to $F^{-1}M =: \tilde{M}$ as required because

$$R = \bigsqcup_{i=1}^{3} (F_T^i|)^{-1} \circ S \circ F_T \colon \tilde{T} \to \tilde{T}_{(2,0).(1,0)^2} = F_T^{-1}(T_{(2,0).(1,0)^2})$$

is a smooth retraction as well. Furthermore, supposing that $\tilde{T}$ is sufficiently small, F_M is stable by [dP, Proposition 2.4] since M is transverse to the presentation $\left(E_{2,0}(\lambda).E^2_{1,0}\right)_{\text{strict}}(F_T)$.

This construction concludes the proof of Proposition 137. □

We note that the tubular neighborhood retraction $T \to I_{(2,0).(1,0)^2}$ restricts to a retraction $M \to I_{(2,0).(1,0)^2}$, giving M a tubular neighborhood structure, and similarly in source; furthermore we can trivialize F_M over T_M just like we trivialized F_T in Lemma 135.

We conclude:

Corollary 142. *The restriction F_M can be trivialized over the tubular neighborhoods $(\tilde{T}_M, \tilde{\pi}_M)$ and (T_M, π_M) of $H_{(2,0).(1,0)^2}$ and $I_{(2,0).(1,0)^2}$ in $\tilde{M}$ and M, and the fiber map is stable; in particular it is $\mathscr{A}$-equivalent to $F_{(2,0).(1,0)^2}$. More precisely (writing $\tilde{\Delta} = (2,0).(1,0)^2$):*

(143)

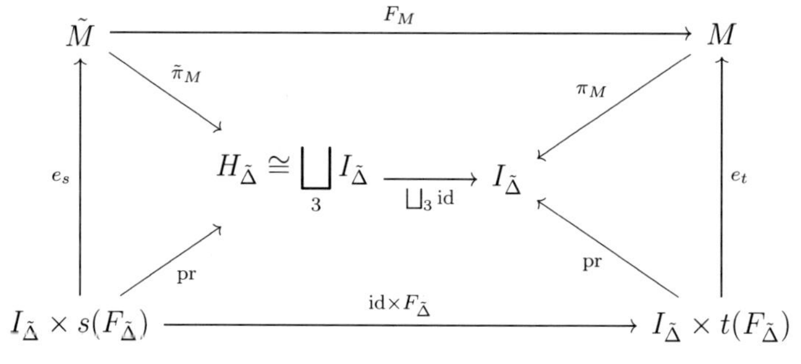

where

$$e = (e_s, e_t)\colon I_{\tilde{\Delta}} \times (s(F_{\tilde{\Delta}}), t(F_{\tilde{\Delta}})) \to (\tilde{M}, M)$$

is a tubular neighborhood embedding. □

Local retractions at the 0-dimensional strata

Yet again, we use Δ to denote either $(2,0)^2$ or $(3,0).(1,0)$. As in Lemma 29 we can define an E-tame retraction

$$(r_\Delta, s_\Delta)\colon F_\Delta \to F_\Delta^+,$$

which we can pull back to a retraction $(\bar{r}_\Delta, \bar{s}_\Delta)$ in (V_Δ, W_Δ) using $(\psi_\Delta, \varphi_\Delta)$:

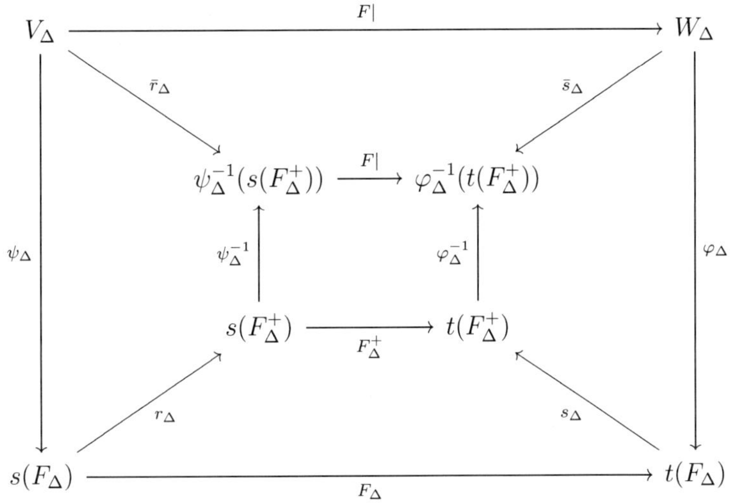

Let us take a closer look at the retraction near I_Δ. Its restriction to the neighborhoods $(\psi_\Delta(\tilde{T}), \varphi_\Delta(T))$ factors as a composition

$$F_{(p_1,0).(p_2,0)} \overset{(r_1,s_1)}{\longrightarrow} F_M \overset{(r_2,s_2)}{\longrightarrow} F^+_{(p_1,0).(p_2,0)}.$$

More precisely, near $I_{(2,0)^2}$ we get:

$$\begin{array}{c}
F_{(2,0)^2} = F_{(2,0)} \times \mathrm{id}_{t(F_{2,0)})} \bigsqcup \mathrm{id}_{t(F_{(2,0)})} \times F_{(2,0)} \\
\Big\downarrow {\scriptstyle (s_1,r_1)} \\
F_M = F_{(2,0)} \times \mathrm{id}_{t(F^+_{(2,0)})} \bigsqcup \mathrm{id}_{t(F_{(2,0)})} \times F^+_{(2,0)} \\
\Big\downarrow {\scriptstyle (r_2,s_2)} \\
F^+_{(2,0)^2} = F^+_{(2,0)} \times \mathrm{id}_{t(F^+_{(2,0)})} \bigsqcup \mathrm{id}_{t(F^+_{(2,0)})} \times F^+_{(2,0)}
\end{array}$$

where

$$(s_1, r_1) = \left(\mathrm{id}_{t(F_{(2,0)})} \times s_{(2,0)} \bigsqcup \mathrm{id}_{t(F_{(2,0)})} \times r_{(2,0)}, \mathrm{id}_{t(F_{(2,0)})} \times s_{(2,0)}\right),$$
$$(r_2, s_2) = \left(r_{(2,0)} \times \mathrm{id}_{t(F^+_{(2,0)})} \bigsqcup s_{(2,0)} \times \mathrm{id}_{t(F^+_{(2,0)})}, s_{(2,0)} \times \mathrm{id}_{t(F^+_{(2,0)})}\right),$$

and near $I_{(3,0).(1,0)}$ we get:

$$F_{(3,0).(1,0)} = F_{(3,0)} \times \mathrm{id}_{t(F_{1,0)})} \bigsqcup \mathrm{id}_{t(F_{(3,0)})} \times F_{(1,0)}$$
$$\Big\downarrow (s_1, r_1)$$
$$F_M = F^1_{(3,0)} \times \mathrm{id}_{t(F_{(1,0)})} \bigsqcup \mathrm{id}_{t(F^1_{(3,0)})} \times F_{(1,0)}$$
$$\Big\downarrow (s_2, r_2)$$
$$F^+_{(3,0).(1,0)} = F^+_{(3,0)} \times \mathrm{id}_{t(F_{(1,0)})} \bigsqcup \mathrm{id}_{t(F^+_{(3,0)})} \times F_{(1,0)}$$

where

$$(s_1, r_1) = \left(r^1_{(3,0)} \times \mathrm{id}_{t(F_{(1,0)})} \bigsqcup s^1_{(3,0)} \times \mathrm{id}_{t(F_{(1,0)})}, s^1_{(3,0)} \times \mathrm{id}_{t(F_{(1,0)})}\right),$$
$$(s_2, r_2) = \left(r^2_{(3,0)} \times \mathrm{id}_{t(F_{(1,0)})} \bigsqcup s^2_{(3,0)} \times \mathrm{id}_{t(F_{(1,0)})}, s^2_{(3,0)} \times \mathrm{id}_{t(F_{(1,0)})}\right).$$

In both cases, (r_1, s_1) is the restriction of an E-tame retraction which is smooth off the negatively weighted axes in the first factor of source and target. These negatively weighted axes lie outside our tubular neighborhoods, so (r_1, s_1) is smooth in $(\psi_\Delta(\tilde{T}), \varphi_\Delta(T))$. Both second retractions (r_2, s_2) are E-tame by Lemma 29.

We need to discuss the behavior of the restrictions of the retractions (r_1, s_1) and (r_2, s_2) to the level sets $\left(F^{-1}\rho_\Delta^{-1}(\epsilon), \rho_\Delta^{-1}(\epsilon)\right)$. As we have already noted, the retraction (r_1, s_1) is smooth, and furthermore it is tangent to the level sets, hence its restriction is just a nice, smooth retraction.

The restriction of the second retraction (r_2, s_2) is just

$$\left(r_{(2,0)}|F^{-1}_{(2,0)}(\rho^{-1}_{(2,0)}(\epsilon)) \times \mathrm{id}_{t(F^+_{(2,0)})} \bigsqcup s_{(2,0)}|\rho^{-1}_{(2,0)}(\epsilon) \times \mathrm{id}_{t(F^+_{(2,0)})},\right.$$
$$\left. s_{(2,0)}|\rho^{-1}_{(2,0)}(\epsilon) \times \mathrm{id}_{t(F^+_{(2,0)})}\right),$$

in the $\Delta = (2,0)^2$ case, and (r_2, s_2) restricts to

$$\left(r^2_{(3,0)}|F^{-1}_{(3,0)}(\rho^{-1}_{(3,0)}(\epsilon)) \times \mathrm{id}_{t(F_{(1,0)})} \bigsqcup s^2_{(3,0)}|\rho^{-1}_{(3,0)}(\epsilon) \times \mathrm{id}_{t(F_{(1,0)})},\right.$$
$$\left. s^2_{(3,0)}|\rho^{-1}_{(3,0)}(\epsilon) \times \mathrm{id}_{t(F_{(1,0)})}\right),$$

in the $\Delta = (3,0).(1,0)$ case. Recall the construction of $(r_{(2,0)}, s_{(2,0)})$ and $(r_{(3,0)}, s_{(3,0)})$. The germs of their restrictions to $(F^{-1}_{(2,0)}\rho^{-1}_{(2,0)}(\epsilon), \rho^{-1}_{(2,0)}(\epsilon))$ and $(F^{-1}_{(3,0)}\rho^{-1}_{(3,0)}(\epsilon), \rho^{-1}_{(3,0)}(\epsilon))$, combined with $(r_{(2,0)}, s_{(2,0)})$ and $(r_{(1,0)}, s_{(1,0)})$ in source and target of $F_{(2,0)^2}$ and $F_{(3,0).(1,0)}$, respectively, at a point at which an $F_{2,0}.F^2_{1,0}$-singularity is presented, comes from a retraction $(r_{(2,0).(1,0)^2}, s_{(2,0).(1,0)^2})$ in an appropriate set of coordinates (with $\Delta = (3,0).(1,0)$ or $(2,0)^2$ and $\tilde{\Delta} = (2,0).(1,0)^2$):

$$(\psi'_\Delta, \varphi'_\Delta)\colon (F_\Delta^{-1}\rho_\Delta^{-1}(\epsilon), \rho_\Delta^{-1}(\epsilon)) \cap (s(F^1_\Delta), t(F^1_\Delta)) \to (s(F_{\tilde{\Delta}}), t(F_{\tilde{\Delta}}))$$

as in

$$\begin{array}{ccc} F_\Delta^{-1}\rho_\Delta^{-1}(\epsilon) \cap s(F^1_\Delta) & \xrightarrow{F^1_\Delta} & \rho_\Delta^{-1}(\epsilon) \cap t(F^1_\Delta) \\ \Big\downarrow{\scriptstyle \psi'_\Delta} & & \Big\downarrow{\scriptstyle \varphi'_\Delta} \\ s(F_{\tilde{\Delta}}) & \xrightarrow[F_{\tilde{\Delta}}]{} & t(F_{\tilde{\Delta}}) \end{array} \tag{144}$$

The retraction $(r_{(2,0).(1,0)^2}, s_{(2,0).(1,0)^2})$ in the level set extends to

$$(\psi_\Delta(\tilde{T}), \phi_\Delta(T)) \cap \left(F_\Delta^{-1}\rho_\Delta^{-1}(0,\epsilon], \rho_\Delta^{-1}(0,\epsilon]\right)$$

using the $\mathbb{R}^+$-action. This retraction pulls back by $(\psi_\Delta, \varphi_\Delta)$ to a retraction $(\tilde{r}_\Delta, \tilde{s}_\Delta)$ in $(\tilde{M}, M)$, leaving the fibers of $(\tilde{\pi}_M, \pi_M)$ invariant.

Lemma 145. *The total local retraction $(r_{\tilde{y}}, s_{\tilde{y}})\colon (F_{(4,0)})_\epsilon \to F \to F^+_\Delta$ at $\tilde{y}$ actually induces a local E-tame retraction $(F_{(4,0)})_\epsilon \to (F^+_{(4,0)})_\epsilon =: F^+$ at $\tilde{y}$.*

Proof. The retraction $(r_{\tilde{y}}, s_{\tilde{y}})$ is E-tame, and hence leaves ST-invariant strata invariant. The fibers of $(r_{\tilde{y}}, s_{\tilde{y}})$ are 3-dimensional, and we know that the strict presentation $(E_\Delta)_{\text{strict}}((F_{(4,0)})_\epsilon)$ is a 3-dimensional, ST-invariant stratum. Thus it must be a fiber of $s_{\tilde{y}}$. But by Lemma 124, the stratum $(E_\Delta)_{\text{strict}}(F_{(4,0)})$ projects submersively onto the non-positively weighted subspace U of $t(F_{(4,0)})$, and since

$$\rho_{(4,0)}^{-1}(\epsilon) = \left(\rho_{(4,0)}^{-1}(\epsilon) \cap t(F^+_{(4,0)})\right) \times U,$$

the manifold $E_\Delta((F_{(4,0)})_\epsilon) \subset t(F_{(4,0)})$ projects submersively onto U as well. Hence it is transverse to $t(F^+_{(4,0)})$ in $t(F_{(4,0)})$, and in particular to $t(F^+)$ in $t((F_{(4,0)})_\epsilon)$. Since transversality is an open property, this means that the fibers of $s_{\tilde{y}}$ are transverse to $t(F^+)$ near $\tilde{y}$, and $s_{\tilde{y}}$ defines a local retraction onto F^+ at $\tilde{y}$. □

Local retractions near the 1-dimensional stratum

As in Lemma 29 we construct a retraction

$$(r_{(2,0).(1,0)^2}, s_{(2,0).(1,0)^2})\colon F_{(2,0).(1,0)^2} := F_{(2,0).(1,0)^2} \to F^+_{(2,0).(1,0)^2}.$$

Using the trivialization

$$e\colon (s(F_{\tilde{\Delta}}) \times I_{\tilde{\Delta}}, t(F_{\tilde{\Delta}}) \times I_{\tilde{\Delta}}) \xrightarrow{\cong} (\tilde{M}, M) \tag{146}$$

where $\tilde{\Delta} = (2,0).(1,0)^2$, we may extend it to a retraction

$$(r_M, s_M)\colon F_M \to F|e\left(s(F^+_{\tilde{\Delta}}) \times I_{\tilde{\Delta}}, t(F^+_{\tilde{\Delta}}) \times I_{\tilde{\Delta}}\right) =: \tilde{F}_{\tilde{\Delta}},$$

where $\tilde{\Delta} = (2,0).(1,0)^2$. In particular, this defines a retraction

$$(147) \qquad \begin{array}{c}(r_{L_y}, s_{L_y})\colon \\ F_M|(\tilde{\pi}_M^{-1}(F^{-1}(y)\cap \Sigma F), \pi_M^{-1}(y)) \to F|e(s(F_{\tilde{\Delta}}) \times \{y\}, t(F_{\tilde{\Delta}}) \times \{y\}).\end{array}$$

Lemma 148. *Suppose that $y \in I_{(2,0).(1,0)^2}$. We claim, as in the previous section, that the total local retraction*

$$(r_y, s_y)\colon (F_{(4,0)})_\epsilon \to F_M \overset{(r_M,s_M)}{\longrightarrow} \tilde{F}_{(2,0).(1,0)^2}$$

induces a local E-tame retraction $(F_{(4,0)})_\epsilon \to (F^+_{(4,0)})_\epsilon$ at y.

Proof. The retraction (r_y, s_y) is E-tame, hence leaves ST-invariant strata invariant. The stratum made up by the presentation $E_{2,0}.E^2_{1,0}((F_{(4,0)})_\epsilon)$ is an ST-invariant stratum, containing the fibers of s_y *and* the germ at y of the set $I_{(2,0).(1,0)^2}$.

By E-tameness, the fiber $s_y^{-1}(y)$ is transverse to

$$e(t(F^+_{(2,0).(1,0)^2}) \times I_{(2,0).(1,0)^2}),$$

but it is also transverse to $I_{(2,0).(1,0)^2}$ in $(E_{2,0}.E^2_{1,0})_{\text{strict}}((F_{(4,0)})_\epsilon)$. That is,

$$T_y\left(E_{2,0}.E^2_{1,0}((F_{(4,0)})_\epsilon)\right) = T_y\left(I_{(2,0).(1,0)^2}\right) \oplus T_y s^{-1}(y)$$

and we know by Lemma 124 that

$$T_y\left(t((F_{(4,0)})_\epsilon)\right) = T_y\left((E_{2,0}.E^2_{1,0})_{\text{strict}}((F_{(4,0)})_\epsilon)\right) \oplus T_y\left(t((F^+_{(4,0)})_\epsilon)\right),$$

so

$$\begin{aligned} T_y\left(t((F_{(4,0)})_\epsilon)\right) &= T_y s^{-1}(y) \oplus \underbrace{T_y\left(I_{(2,0).(1,0)^2}\right) \oplus T_y\left(t((F^+_{(4,0)})_\epsilon)\right)}_{=T_y\left(t((F^+_{(4,0)})_\epsilon)\right)} \\ &= T_y s^{-1}(y) \oplus T_y\left(t((F^+_{(4,0)})_\epsilon)\right)\end{aligned}$$

so the fiber $s_y^{-1}(y)$ is transverse to $t((F^+_{(4,0)})_\epsilon)$ in $t((F_{(4,0)})_\epsilon)$, and hence s_y defines a local E-tame retraction $(F_{(4,0)})_\epsilon \to (F^+_{(4,0)})_\epsilon$. □

Local retractions near $I(F^+)$

We start out by recalling that for $y \in I_{(2,0).(1,0)^2} \cap W_\Delta$, where, as before, $\Delta \in \{(2,0)^2, (3,0).(1,0)\}$, we have two sets of natural diffeomorphisms

$$\begin{aligned}\varphi_a, \varphi_b &\colon (\pi_M^{-1}(y), y) \to (t(F_{(2,0).(1,0)^2}), 0),\\ \psi_a, \psi_b &\colon (\tilde{\pi}_M^{-1}(F^{-1}(y)\cap \Sigma F), y) \to (s(F_{(2,0).(1,0)^2}), 0).\end{aligned}$$

Both diffeomorphism-pairs define weighted homogeneous coordinates at y in the fibers, where

$$\begin{array}{c}(\psi_a, \varphi_a)_y = (\psi'_\Delta, \varphi'_\Delta) \circ (\psi_\Delta, \varphi_\Delta)\colon \\ (\tilde{\pi}_M^{-1}(F^{-1}(y)\cap \Sigma F), \pi_M^{-1}(y)) \to (F_\Delta^{-1}\rho_\Delta^{-1}(\epsilon), \rho_\Delta^{-1}(\epsilon)) \to \\ (s(F_{(2,0).(1,0)^2}), t(F_{(2,0).(1,0)^2}))\end{array}$$

is the diffeomorphism that brings the retraction $(\tilde{s}_\Delta, \tilde{r}_\Delta)$ defined in (4.1.6) to

$$(r_{(2,0).(1,0)^2}, s_{(2,0).(1,0)^2}),$$

and where

$$(\psi_b, \varphi_b)_y := e^{-1}|\colon$$
$$(\tilde{\pi}_M^{-1}(F^{-1}(y) \cap \Sigma F), \pi_M^{-1}(y)) \to (s(F_{(2,0).(1,0)^2}) \times \{y\}, t(F_{(2,0).(1,0)^2}) \times \{t\})$$

is the inverse of the map e defined in (146), which takes the fiber retraction (r_{L_y}, s_{L_y}) defined in (147) to $(r_{(2,0).(1,0)^2}, s_{(2,0).(1,0)^2})$.

Suppose that

$$\alpha\colon]0,1[\to I_{(2,0).(1,0)^2}$$

is a smooth embedding of $]0,1[$ onto one of the components of $I_{(2,0).(1,0)^2}$, and that α has an extension $\bar{\alpha}\colon I \to I_{(2,0).(1,0)^2}$ such that $\bar{\alpha}(0)$ and $\bar{\alpha}(1)$ are components of $I_{(2,0)^2}$ or $I_{(3,0).(1,0)}$. Together with the trivialization (143) this defines a commutative diagram (and we agree to denote $\tilde{\Delta} = (2,0).(1,0)^2$ in the diagrams):

$$\begin{array}{ccc}
]0,1[\times s(F_{\tilde{\Delta}}) & \xrightarrow{\mathrm{id}\times F_{\tilde{\Delta}}} &]0,1[\times t(F_{\tilde{\Delta}}) \\
\downarrow{\scriptstyle \alpha\times\mathrm{id}} & & \downarrow{\scriptstyle \alpha\times\mathrm{id}} \\
I_{\tilde{\Delta}} \times s(F_{\tilde{\Delta}}) & \xrightarrow[\mathrm{id}\times F_{\tilde{\Delta}}]{} & I_{\tilde{\Delta}} \times t(F_{\tilde{\Delta}}) \\
\downarrow{\scriptstyle e_s} & & \downarrow{\scriptstyle e_t} \\
\tilde{M} & \xrightarrow[F_M]{} & M
\end{array}$$

Denote by (ψ_1, ϕ_1) the vertically composed diffeomorphisms.

On the other hand, we recall from p. 91 that there are two other decompositions, namely near $\bar{\alpha}(0)$:

$$\begin{array}{ccc}
\tilde{M} \cap \tilde{V}_{\bar{\alpha}(0)} & \xrightarrow{F_M|} & M \cap \tilde{W}_{\bar{\alpha}(0)} \\
\downarrow{\scriptstyle \psi_{\bar{\alpha}(0)}} & & \downarrow{\scriptstyle \phi_{\bar{\alpha}(0)}} \\
s(F^1_{\bar{\alpha}(0)}) \cap \psi_{\bar{\alpha}(0)}(\tilde{T}) & \xrightarrow[F^1_{\bar{\alpha}(0)}|]{} & t(F^1_{\bar{\alpha}(0)}) \cap \phi_{\bar{\alpha}(0)}(T) \\
\downarrow & & \downarrow \\
]0,\epsilon[\times s(F_{\tilde{\Delta}}) & \xrightarrow[\mathrm{id}\times F_{\tilde{\Delta}}]{} &]0,\epsilon[\times t(F_{\tilde{\Delta}})
\end{array}$$

where we denote by (ψ_2, ϕ_2) the vertically composed diffeomorphisms, and near $\bar{\alpha}(1)$:

$$\begin{array}{ccc}
\tilde{M} \cap \tilde{V}_{\bar{\alpha}(1)} & \xrightarrow{F_M|} & M \cap \tilde{W}_{\bar{\alpha}(1)} \\
\Big\downarrow{\scriptstyle \psi_{\bar{\alpha}(1)}} & & \Big\downarrow{\scriptstyle \phi_{\bar{\alpha}(1)}} \\
s(F^1_{\bar{\alpha}(1)}) \cap \psi_{\bar{\alpha}(1)}(\tilde{T}) & \xrightarrow{F^1_{\bar{\alpha}(1)}|} & t(F^1_{\bar{\alpha}(1)}) \cap \phi_{\bar{\alpha}(1)}(T) \\
\Big\downarrow & & \Big\downarrow \\
]1-\epsilon, 1[\times s(F_{\tilde{\Delta}}) & \xrightarrow[\mathrm{id}\times F_{\tilde{\Delta}}]{} &]1-\epsilon, 1[\times t(F_{\tilde{\Delta}})
\end{array}$$

Here we denote the vertically composed diffeomorphisms by (ψ_3, ϕ_3). Note that (ψ_1, ϕ_1) together with (ψ_2, ϕ_2) or (ψ_3, ϕ_3) define the fiber diffeomorphism pairs (ψ_a, φ_a) and (ψ_b, φ_b) defined above.

Now the composed diffeomorphisms $\psi_2 \circ \psi_1$ and $\phi_2 \circ \phi_1$ define a commutative diagram

$$\begin{array}{ccc}
]0, \epsilon[\times s(F_{\tilde{\Delta}}) & \xrightarrow{\mathrm{id}\times F_{\tilde{\Delta}}} &]0, \epsilon[\times t(F_{\tilde{\Delta}}) \\
\Big\downarrow{\scriptstyle \psi_2 \circ \psi_1|} & & \Big\downarrow{\scriptstyle \phi_2 \circ \phi_1|} \\
]0, \epsilon[\times s(F_{\tilde{\Delta}}) & \xrightarrow[\mathrm{id}\times F_{\tilde{\Delta}}]{} &]0, \epsilon[\times t(F_{\tilde{\Delta}})
\end{array}$$

where the maps $\psi_3 \circ \psi_1$ and $\phi_3 \circ \phi_1$ are fibered over $]0, \epsilon[$, and similarly

$$\begin{array}{ccc}
]1-\epsilon, 1[\times s(F_{\tilde{\Delta}}) & \xrightarrow{\mathrm{id}\times F_{\tilde{\Delta}}} &]1-\epsilon, 1[\times t(F_{\tilde{\Delta}}) \\
\Big\downarrow{\scriptstyle \psi_3 \circ \psi_1|} & & \Big\downarrow{\scriptstyle \phi_3 \circ \phi_1|} \\
]1-\epsilon, 1[\times s(F_{\tilde{\Delta}}) & \xrightarrow[\mathrm{id}\times F_{\tilde{\Delta}}]{} &]1-\epsilon, 1[\times t(F_{\tilde{\Delta}})
\end{array}$$

Using these fibered diffeomorphisms, we obtain smooth paths

$$\gamma_1 \colon]0, \epsilon[\to \mathscr{A}_{F_{(2,0).(1,0)^2}} \text{ and } \gamma_2 \colon]1-\epsilon, 1[\to \mathscr{A}_{F_{(2,0).(1,0)^2}},$$

such that

$$\begin{aligned}
\gamma_1(t) \cdot (\psi_b, \varphi_b)_{\bar{\alpha}(t)} &= (\psi_a, \varphi_a)_{\bar{\alpha}(t)}, \quad t \in [0, \epsilon], \\
\gamma_2(t) \cdot (\psi_b, \varphi_b)_{\bar{\alpha}(t)} &= (\psi_a, \varphi_a)_{\bar{\alpha}(t)}, \quad t \in [1-\epsilon, 1],
\end{aligned}$$

as shown schematically in the following diagram, still with $\tilde{\Delta} = (2,0).(1,0)^2$:

$$
\begin{array}{ccccccc}
\{t\} \times s(F_{\tilde{\Delta}}) & & & \xrightarrow{\quad F_{\tilde{\Delta}} \quad} & & & \{t\} \times t(F_{\tilde{\Delta}}) \\
 & \nwarrow{\scriptstyle \psi_{L_M}} & & & & \nearrow{\scriptstyle \varphi_{L_M}} & \\
\Big\downarrow{\scriptstyle \psi_2 \circ \psi_1} & & \tilde{\pi}_M^{-1} F_M^{-1}(y)) & \xrightarrow{F_M|} & \pi_M^{-1}(y) & & \Big\downarrow{\scriptstyle \varphi_2 \circ \varphi_1} \\
 & \swarrow{\scriptstyle \psi_\epsilon \circ \psi_{(2,0)^2}} & & & & \searrow{\scriptstyle \varphi_\epsilon \circ \varphi_\Delta} & \\
\{t\} \times s(F_{\tilde{\Delta}}) & & & \xrightarrow{\quad F_{\tilde{\Delta}} \quad} & & & \{t\} \times t(F_{\tilde{\Delta}})
\end{array}
$$

when $t < \epsilon$ and $\rho_\Delta(y) = t$, and similarly when $t > 1 - \epsilon$.

Let $\epsilon' < \epsilon$, and fix a maximal compact subgroup G of $\mathscr{A}_{F_{(2,0).(1,0)^2}}$. By Theorem 97 we may assume that $G = \{\pm 1\} \times \{\pm 1\} \times \{\pm 1\} < \mathscr{A}_{F_{(2,0)}} \times \mathscr{A}_{F_{(1,0)}} \times \mathscr{A}_{F_{(1,0)}}$. By Corollary 85, there exists a smooth path $\gamma_3 \colon [\epsilon', \epsilon] \to \mathscr{A}_{F_{(2,0).(1,0)^2}}$ such that $\gamma_3(\epsilon') = \beta_0 \cdot \gamma_1(\epsilon')$ for some element $\beta_0 \in MC(\mathscr{A}_{F_{(2,0).(1,0)^2}})$, and $\gamma_3(\epsilon) = \mathrm{id}$. Similarly there exist a smooth path $\gamma_4 \colon [1-\epsilon, 1-\epsilon'] \to \mathscr{A}_{F_{(2,0).(1,0)^2}}$ and an element $\beta_1 \in \mathscr{A}_{F_{(2,0).(1,0)^2}}$ such that $\gamma_4(1-\epsilon) = \mathrm{id}$ and $\gamma_4(1-\epsilon') = \beta_1 \cdot \gamma_2(1-\epsilon')$. Define a smooth path $\gamma_5 \colon [\epsilon', 1-\epsilon'] \to \mathscr{A}_{F_{(2,0).(1,0)^2}}$ by setting

$$
\begin{aligned}
&\gamma_5|[\epsilon', \epsilon] = \gamma_3 \circ \kappa_1[\epsilon', \epsilon] \\
&\gamma_5|[\epsilon, 1-\epsilon] = \mathrm{id} \\
&\gamma_5|[1-\epsilon, 1] = \gamma_4 \circ \kappa_2[1-\epsilon, 1-\epsilon']
\end{aligned}
$$

where $\kappa_1 \colon [\epsilon', \epsilon] \to [\epsilon', \epsilon]$ and $\kappa_2 \colon [1-\epsilon, 1-\epsilon'] \to [1-\epsilon, 1-\epsilon']$ are smooth reparametrizations such that the nth derivatives $\kappa_1^{(n)}(\epsilon') = \kappa_1^{(n)}(\epsilon) = \kappa_2^{(n)}(1-\epsilon) = \kappa_2^{(n)}(1-\epsilon') = 0$ for all $n \in \mathbb{N}$ (this is to ensure that γ_5 is smooth).

Now we have commutative diagrams

(149)

$$
\begin{array}{ccccccc}
\tilde{M} \cap \tilde{V}_{\tilde{\alpha}(0)} & & & \xrightarrow{\quad F_M| \quad} & & & M \cap \tilde{W}_{\tilde{\alpha}(0)} \\
 & \searrow{\scriptstyle \psi_1^{-1}} & & & & \swarrow{\scriptstyle \phi_1^{-1}} & \\
\Big\downarrow{\scriptstyle \psi_2} & &]0, \epsilon'[\times s(F_{\tilde{\Delta}}) & \xrightarrow{\mathrm{id} \times F_{\tilde{\Delta}}} &]0, \epsilon'[\times t(F_{\tilde{\Delta}}) & & \Big\downarrow{\scriptstyle \phi_2} \\
 & \swarrow{\scriptstyle (t,x) \mapsto (t, \gamma_1(t)(x))} & & & & \searrow{\scriptstyle (t,y) \mapsto (t, \gamma_1(t)(y))} & \\
]0, \epsilon'[\times s(F_{\tilde{\Delta}}) & & & \xrightarrow[\mathrm{id} \times F_{\tilde{\Delta}}]{} & & &]0, \epsilon'[\times t(F_{\tilde{\Delta}})
\end{array}
$$

and – setting $\delta = 1 - \epsilon'$,

$$(150)\qquad \begin{array}{ccccccc}
\tilde{M} \cap \tilde{V}_{\bar{\alpha}(1)} & & & \xrightarrow{F_M|} & & & M \cap \tilde{W}_{\bar{\alpha}(1)} \\
\downarrow{\scriptstyle \psi_2} & \searrow^{\psi_1^{-1}} & & & & \swarrow^{\phi_1^{-1}} & \downarrow{\scriptstyle \phi_2} \\
& &]\delta,1[\times s(F_{\tilde{\Delta}}) & \xrightarrow{\mathrm{id}\times F_{\tilde{\Delta}}} &]\delta,1[\times t(F_{\tilde{\Delta}}) & & \\
& \swarrow_{(t,x)\mapsto(t,\gamma_2(t)(x))} & & & & \searrow_{(t,y)\mapsto(t,\gamma_2(t)(y))} & \\
]\delta,1[\times s(F_{\tilde{\Delta}}) & & & \xrightarrow[\mathrm{id}\times F_{\tilde{\Delta}}]{} & & &]\delta,1[\times t(F_{\tilde{\Delta}})
\end{array}$$

and

$$(151)\qquad \begin{array}{ccc}
\tilde{\pi}_M^{-1}(\alpha^{-1}(]\epsilon',\delta[)) & \xrightarrow{F_M|} & \pi_M^{-1}(\alpha^{-1}(]\epsilon',\delta[)) \\
\downarrow{\scriptstyle \psi_1^{-1}} & & \downarrow{\scriptstyle \phi_1^{-1}} \\
]\epsilon',\delta[\times s(F_{\tilde{\Delta}}) & \xrightarrow{\mathrm{id}\times F_{\tilde{\Delta}}} &]\epsilon',\delta[\times t(F_{\tilde{\Delta}}) \\
\downarrow{\scriptstyle (t,x)\mapsto(t,\gamma_5(t)(x))} & & \downarrow{\scriptstyle (t,y)\mapsto(t,\gamma_5(t)(y))} \\
]\epsilon',\delta[\times s(F_{\tilde{\Delta}}) & \xrightarrow[\mathrm{id}\times F_{\tilde{\Delta}}]{} &]\epsilon',\delta[\times t(F_{\tilde{\Delta}})
\end{array}$$

Here

$$\gamma_5(\epsilon') = \gamma_3(\epsilon') = \beta_0 \cdot \gamma_1(\epsilon'),$$
$$\gamma_5(\delta) = \gamma_4(\delta) = \beta_1 \cdot \gamma_2(\delta),$$

where β_0 and β_1 are of the form $\{\pm 1\}\times\{\pm 1\}\times\{\pm 1\}$, acting on $F_{(2,0).(1,0)^2}$ through the weighted homogeneous $\mathbb{R}^*$-action on the factors in source and target. In particular, the two choices of subspaces $(s(F^+_{(2,0).(1,0)^2}), t(F^+_{(2,0).(1,0)^2}))$ coincide.

We pull the retractions $(r_{(2,0).(1,0)^2}, s_{(2,0).(1,0)^2})$ back fiberwise using the vertical diffeomorphisms, and note that we get two choices of retractions in the fibers at $\bar{\alpha}(\epsilon')$ and $\bar{\alpha}(\delta)$ differing by an element of $MC(\mathscr{A}_{F_{(2,0).(1,0)^2}})$. But by the results in Chapter 4.1.4, the set of fibers of $(r_{(2,0)}, s_{(2,0)})$, is invariant with respect to $\{\pm 1\}$, as is the set of fibers of the $(r_{(1,0)}, s_{(1,0)})$, which are just identities. Hence the set of fibers of $(r_{(2,0).(1,0)^2}, s_{(2,0).(1,0)^2})$ is $\{\pm 1\}\times\{\pm 1\}\times\{\pm 1\}$-invariant as well. It follows that the fibers of the retractions defined by pulling the retraction $(r_{(2,0).(1,0)^2}, s_{(2,0).(1,0)^2})$ back using the vertical diffeomorphisms in (149) and (150) coincide – at $\bar{\alpha}(\epsilon')$ and $\bar{\alpha}(\delta)$ – with those defined by pulling back using the vertical diffeomorphisms in (151). Thus (possibly using reparametrizations of $[0, \epsilon']$ and $[1-\epsilon', 1]$ which are flat at ϵ' and $1-\epsilon'$), we get a well-defined $C^{0,1}$-foliation transverse to any possible choice of $t(F^+_{(2,0).(1,0)^2})$ in the fibers. Hence we get a well-defined retraction $F_{(2,0).(1,0)^2} \to F^+_{(2,0).(1,0)^2}$ in the fibers which coincides with the one defined in (144) near I_Δ.

We note that by the same reasoning as in Lemma 145, all of these coinciding retractions induce, in fact, an E-tame retraction $(R_1, S_1)\colon (F_{(4,0)})_\epsilon \to (F^+_{(4,0)})_\epsilon$ in

$(F^{-1}W, W)$, where W is a neighborhood of the instability locus.

Construction of the global retraction

To finish the construction we combine the retractions (R_1, S_1) with smooth retractions (R_2, S_2) off the positive instability locus and obtain an E-tame retraction $(R, S)\colon F_\epsilon \to F_\epsilon^+$ using Lemma 32, with the help of a submersive distance function $h\colon W \to \mathbb{R}$, which fibers $h^{-1}(]0,\infty[)$ over $]0,\infty[$, and which pulls back to a fibration over $]0,\infty[$ in source as well. For instance, we can use the weighted homogeneous distance function in the fibers of T, combined with the local weighted homogeneous distance function near the 0-dimensional strata $I_{(2,0)^2}$ and $I_{(3,0).(1,0)}$; it will pull back nicely to $s(F_\epsilon)$ because, being transverse to orbits of the local $\mathbb{R}^+$-action, its level sets will be transverse to F_ϵ.

We have proven:

Theorem 152. *There exists an E-tame retraction*

$$(r_{(4,0)}, s_{(4,0)})\colon F_{(4,0)} \to F_{(4,0)}^+,$$

and $F_{(4,0)}^+$ is topologically ministable. □

4.2. **Z-singularities.** The singularities belonging to the series $Z_{p,0}(*)$ have many properties in common with the $E_{p,0}(*)$-singularities, for instance when it comes to the geometry of the instability locus and the germs presented there. We can extend our results for the E-series to the Z-series using the same techniques, and this section is dedicated to showing how this can be done. We will dwell on differences rather than similarities, and simply refer to the proof for the E-series case when the ideas are the same.

A function $\mathbb{R}^2 \to \mathbb{R}$ whose 4-jet is xy^3 belongs to the Z^1-series. Following [dPW04] we call it the Z-series, as no other Z-singularities will appear in this thesis. We concentrate on the singularity $Z_{p,0}(*)$, whose normal forms for $\mathscr{R}$-equivalence are given by

$$x(y^3 + yx^{2p}W_p + x^{3p}),$$

for $p \geq 1$, where

$$W_p = w_0 + w_1 x + \ldots + w_{p-1}x^{p-1}$$

as before, and where we assume $4w_0^3 + 27 \neq 0$ in order to get finitely $\mathscr{K}$-determined germs, and $w_0 \neq 0$ because – as in the E-series case – the germs $x(y^3 + x^{3p})$ and $x(y^3 + yx^{2p})$ give rise to deformations which are topologically distinct from the deformations found in the other cases [dPW04, Theorem 6.3].

For $\mathscr{K}$-equivalence we get normal forms $Z_{p,0}(r)$ given by

$$x(y^3 + yx^{2p}(w_0 + x^r) + x^{3p}), \tag{153}$$

where $r = \min\{i > 0 | w_i \neq 0\}$.

We shall focus on the case where $r = p$, which is $\mathscr{K}$-equivalent to the weighted homogeneous germ

$$x(y^3 + \lambda y x^{2p} + x^{3p}), \tag{154}$$

for some $\lambda \neq 0$ with $4\lambda^3 + 27 \neq 0$. Like in the E-series case, if we can construct an E-tame retraction $F_{(p,0)} \to F^+_{(p,0)}$ for the unfoldings of the weighted homogeneous case (154), then we get E-tame retractions $F_{(p,0)} \to F^+_{(p,0)}$ for the unfoldings of the other germs (153) for free. See Section 4.3.

The ministable unfolding of (154) is given by

$$F_{(p,0)}\colon \mathbb{R}^{6p+4} \to \mathbb{R}^{6p+3}, \quad (x, y, \underline{u}, \underline{v}, c) \mapsto$$
$$\left(x(y^3 + \lambda y x^{2p} + x^{3p}) + \textstyle\sum_{i=1}^{3p} u_i x^i + y \sum_{j=0}^{3p} v_j x^j + cy^2, \underline{u}, \underline{v}, c\right)$$

We agree to, in this chapter, denote the standard unfolding for $Z_{p,0}(*)$ by $F_{(p,0)}$, or $F^Z_{(p,0)}$ if we want to emphasize the singularity type Z. The standard unfolding for $E_{p,0}(*)$ will be denoted $F^E_{(p,0)}$.

Fix $\underline{u}$, $\underline{v}$ and c. The germ

$$(x, y) \mapsto \mathrm{pr}_1 \circ F_{(p,0)}(x, y, \underline{u}, \underline{v}, c)$$

lies in $Z_{p,0}(*)$ if and only if $c = 0$, $\underline{u} = 0$, $v_j = 0$ for $j < 2p + 1$ and $v_{2p+1} \neq -\lambda$. In particular, $Z_{p,0}(*)$ is presented on smooth submanifolds of the unfolding parameter spaces of $s(F_{(p,0)})$ and $t(F_{(p,0)})$.

The total target weight of f is $3p + 1$, and we assign source weights

$$\begin{cases} x \rightsquigarrow 1 > 0, \\ y \rightsquigarrow p > 0, \\ v_j \rightsquigarrow 3p + 1 - p - j = 2p + 1 - j > 0 \Leftrightarrow 2p + 1 > j, \\ u_i \rightsquigarrow 3p + 1 - i > 0, \\ c \rightsquigarrow 3p + 1 - 2p = p + 1 > 0, \end{cases}$$

so $F_{(p,0)}$ has p non-positively weighted unfolding variables $v_{2p+1}, \ldots, v_{3p}$, with weights $0, -1, \ldots, -(p-1)$. In particular, $F^+_{(p,0)} = F^p_{(p,0)}$.

Our goal is, of course, to construct an E-tame retraction $F_{(p,0)} \to F^+_{(p,0)}$. Tame retractions have been constructed by du Plessis and Wall for $p = 1, 2$ [dPW95, Chapter 10]. We construct E-tame retractions for $p = 1, 2, 3$.

4.2.1. *Instability loci and presented singularities.*

Proposition 155. *Provided $k >> 0$, the germ class $Z_{p,0}(*)$ defines a smooth submanifold of the jet space $J^k(m, m-1)$, and it has the immersion condition. In particular, given any stable map $F\colon M \to N$, the strict presentation $\Delta(F)$ of any multigerm class containing $Z_{p,0}(*)$ will be a smooth submanifold of $t(F)$, and the restriction*

$$F|\colon F^{-1}(\Delta(F)) \cap \Sigma F \to \Delta(F)$$

restricts to a diffeomorphism on components.

Proof. The proof that $Z_{p,0}(*)$ defines a smooth submanifold of $J^k(m, m+1)$ follows the proof of Proposition 112. The proof that $Z_{p,0}(*)$ has the immersion condition follows the proof of Proposition 115, using the fact that $Z_{p,0}(*)$ is presented on a smooth submanifold of the unfolding parameter subspace of $t(F_{(p,0)})$. The rest follows as in Corollaries 116 and 117. □

The following theorem by du Plessis and Wall gives a parametrization of the instability locus of $F^k_{(p,0)}$, where $k \leq p$:

Theorem 156. [dPW04, Theorem 7.1] *For any $k \leq p$, the instability locus of F^k is the union of the images of the following deformations (with $s \leq k-1$):*

$$x\left(y^3 + \lambda y \prod_{i=*}^{s}(x-\xi_i)^{2c_i} + \prod_{i=*}^{s}(x-\xi_i)^{3c_i}\right), \quad \sum_{i=*}^{s} c_i = p. \tag{157}$$

where $*$ denotes either 0 *or* 1, *and where $\xi_0 \equiv 0$.* □

Remark 158. Note that as in the E-series case, it is enough to consider the deformations

$$\begin{gathered}(x, y, \xi_1, \ldots, \xi_{k-1}) \mapsto \\ x\left(y^3 + \lambda y x^{2c_0}\prod_{i=1}^{k-1}(x-\xi_i)^{2c_i} + x^{3c_0}\prod_{i=1}^{k-1}(x-\xi_i)^{3c_i}\right), \\ \sum_{i=0}^{k-1} c_i = p,\end{gathered} \tag{159}$$

because the $s < k-1$ cases from (157) are found on subspaces of the parameter space of type $\{\xi_i = \xi_j\}$ or $\{\xi_i = 0\}$. The case $k = p$ gives the instability locus of $F^+_{(p,0)}$, and it suffices to consider the restriction of the following deformation D_Z:

$$\begin{gathered}D_Z^+ \colon (x, y, \xi_1, \ldots, \xi_p) \mapsto \\ x\left(y^3 + \lambda y \prod_{i=1}^{p}(x-\xi_i)^2 + \prod_{i=1}^{p}(x-\xi_i)^3\right)\end{gathered}$$

to the subset

$$P = \{(\xi_1, \ldots, \xi_p) \in \mathbb{R}^p | \xi_i = 0 \text{ for some } i \text{ or } \xi_j = \xi_k \text{ for some } j, k\}.$$

We form the parametrization $p_Z \colon P \to t(F^+_{(p,0)})$ of the positive instability locus by setting

$$p_Z = (u_1, \ldots, u_{3p+1}, v_0, \ldots, v_{2p+1}, c)(\xi_1, \ldots, \xi_p),$$

where

$$\begin{aligned} u_i(\xi_1, \ldots, \xi_p) &= \operatorname{coef}(x^i, D_Z^+) \\ v_j(\xi_1, \ldots, \xi_p) &= \operatorname{coef}(yx^j, D_Z^+) \\ c(\xi_1, \ldots, \xi_p) &= 0. \end{aligned}$$

Note that as in the E-series case, the parametrizations are symmetric maps, and to find the instability locus it is enough to restrict to the subset $\{(\xi_i) \in P | \xi_1 \leq \ldots \leq \xi_p\}$. Furthermore, we do not have any dependencies between the ξ_i.

In analogy with the E-series case, we have:

Proposition 160.

i) *There is a stratification of the parameter space P by sets of type $\{\xi_i = \ldots = \xi_j\}$, which induces a stratification of the instability locus via p_Z, and the sets $\{\xi_i = \ldots = \xi_j\}$ have constellations (c_i) of exponents in the deformation (159) associated to them.*

Denote by $(c_i)_{i=0}^s$ the constellation of c_i associated to a point $(\xi_1, \ldots, \xi_p)$ of the parameter space. At $p(\xi_i)$ we find presented a

$$\Delta_{c_0,0}(\lambda).E_{c_1,0}(\lambda).\cdots.E_{c_s,0}(\lambda) \tag{161}$$

-singularity (modulo $\mathscr{E}\mathscr{K}$), where $\Delta_{c_0,0}(\lambda)$ is $Z_{c_0,0}(\lambda)$ whenever $c_0 > 0$, and $\Delta_{0,0}(\lambda)$ is

$$\begin{cases} A_1^3 & \text{if } \lambda < -\sqrt[3]{\frac{27}{4}} \\ A_1 & \text{if } \lambda > -\sqrt[3]{\frac{27}{4}} \end{cases}$$

where we note that the value $\lambda = -\sqrt[3]{\frac{27}{4}}$ is never reached because of our finite $\mathscr{K}$-determinacy condition $4\lambda^3 + 27 \neq 0$. Here A_1 denotes the stable rank 0 singularity $F_{A_1} : \mathbb{R}^2 \to \mathbb{R}$, $(x, y) \mapsto xy$.

We denote by $P_{(c_1,0).\cdots.(c_s,0)}$ the set of all points $(\xi_1, \ldots, \xi_{p-1})$ associated to $(c_i)_{i=1}^s$; these sets form the stratification in the parameter space P.

ii) *The parametrization p_Z restricts to a homeomorphism on components of strata.*

iii) *The strata in the instability locus are smooth, with contractible components, and each stratum I_Δ is actually a union of components of the intersection $t(F^+_{(p,0)}) \cap Y_\Delta$, where Y_Δ is the total strict presentation of the singularity Δ in $t(F_{(p,0)})$.*

Remark 162. Note that, as opposed to the statements in [dPW95, p. 507] and [dPW04], which are made simultaneously over $\mathbb{R}$ and $\mathbb{C}$ we only find *one* A_1 singularity in $\Delta_{0,0}(\lambda)$ when $\lambda > -\sqrt[3]{\frac{27}{4}}$ in the real case.

Proof.

i) We show that we find the singularity (161) presented at $p(\xi_i)$. When $c_0 > 0$, we see this in the same way as in Lemma 123 for the E-series. When $c_0 = 0$ we note that when all $\xi_i \neq 0$, the germ at $(0,0)$ of the deformation (with $\xi_j = \xi_k$ for some j, k)

$$(x, y) \mapsto x\left(y^3 + \lambda y \prod_{i=1}^{p}(x - \xi_i)^2 + \prod_{i=1}^{p}(x - \xi_i)^3\right)$$

is equivalent to $x(y^3 + \lambda y + 1)$. Over the complex numbers, this polynomial factors as $x(y - c_1)(y - c_2)(y - c_3)$ for three distinct $c_i \in \mathbb{C}$, which means that if we consider all our polynomials over $\mathbb{C}$, then we get stable A_1-singularities $(x, y) \mapsto xy$ at points $(x, y) = (0, c_i)$.

We first show that when $\lambda < -\sqrt[3]{\frac{27}{4}}$, all of the roots c_i are real and distinct, and we get three stable A_1-singularities presented by F at $(0, c_i)$ for $1 \le i \le 3$.

The function $\chi(y) = y^3 + \lambda y + 1$ has three real, distinct roots if and only if the equation $\chi'(y) = 0$ has two solutions y_1 and y_2, where $\chi(y_i)$ takes one positive and one negative value for $i = 1, 2$. See Figure 10.

We compute $\chi'(y) = 3y^2 + \lambda = 3(y + \sqrt{\frac{-\lambda}{3}})(y - \sqrt{\frac{-\lambda}{3}})$, and see that $\chi'(y) = 0$ if and only if $y = \pm\sqrt{\frac{-\lambda}{3}}$, which is real if and only if $\lambda \le 0$. We evaluate

$$\begin{cases} \chi(\sqrt{\frac{-\lambda}{3}}) = \frac{2\lambda}{3}\sqrt{\frac{-\lambda}{3}} + 1 & \text{which is negative when } \lambda < -\sqrt[3]{\frac{27}{4}}, \text{ and} \\ \chi(-\sqrt{\frac{-\lambda}{3}}) = -\frac{2\lambda}{3}\sqrt{\frac{-\lambda}{3}} + 1 & \text{which is positive.} \end{cases}$$

and see that the claim holds – we find three distinct, real roots if and only if $\lambda < -\sqrt[3]{\frac{27}{4}}$, and it follows that we get three A_1-singularities presented by $F^Z_{(p,0)}$ at $p_Z(\xi_i)$ if and only if $\lambda < -\sqrt[3]{\frac{27}{4}}$.

When $\lambda > -\sqrt[3]{\frac{27}{4}}$, the complexification of χ has three distinct roots – one real and two complex conjugate, non-real roots. These roots give rise to three A_1-singularities presented by the complexification of $F^Z_{(p,0)}$. As the points at which two of the A_1 singularities are presented, are not in the domain of the real polynomial, which can be seen as a restriction of the complex polynomial, we see that the real part of $F^{\mathbb{C}}_{(A_1^3.E_{c_1,0}.\cdots.E_{c_s,0})}$ here is just $F_{(A_1.E_{c_1,0}.\cdots.E_{c_s,0})} \times \mathrm{id}_{\mathbb{R}^2}$, and since we can show (as in the real E-series case) that the presentation

$$(A_1^3.E_{c_1,0}(\lambda).\cdots.E_{c_s,0}(\lambda))_{\text{strict}}(F^{\mathbb{C}}_{(A_1^3.E_{c_1,0}.\cdots.E_{c_s,0})})$$

lies in $t((F^{\mathbb{C}}_{(p,0)})^+)$, the presentation

$$(A_1.E_{c_1,0}(\lambda).\cdots.E_{c_s,0}(\lambda))_{\text{strict}}(F_{(A_1.E_{c_1,0}.\cdots.E_{c_s,0})} \times \mathrm{id}_{\mathbb{R}^2})$$

will lie in $t(F^+_{(p,0)})$. In particular, the $A_1.E_{c_1,0}(\lambda).\cdots.E_{c_s,0}(\lambda)$ singularities are presented on an $s+2$-dimensional subset of $t((F^Z_{(p,0)})^+)$.

ii) The proof is similar to the E-series proof that p is injective on components of the strata, which is part of the proof of Lemma 126.

iii) The proof follows the proof of Lemmas 126 and 130 from the E-series case.

□

Proposition 163. *Let $\prod E_{c_i,0}(\lambda)$ or $Z_{p,0}(\lambda)$ be a singularity appearing in the positive stability locus $I(F^+_{(p,0)})$. Now, for $\Delta = \prod E_{c_i,0}(*)$ or $\Delta = Z_{p,0}(*)$, the components of Y_Δ passing through $I(F^+_{(p,0)})$ project submersively onto the non-positively weighted subspace of $t(F_{(p,0)})$.*

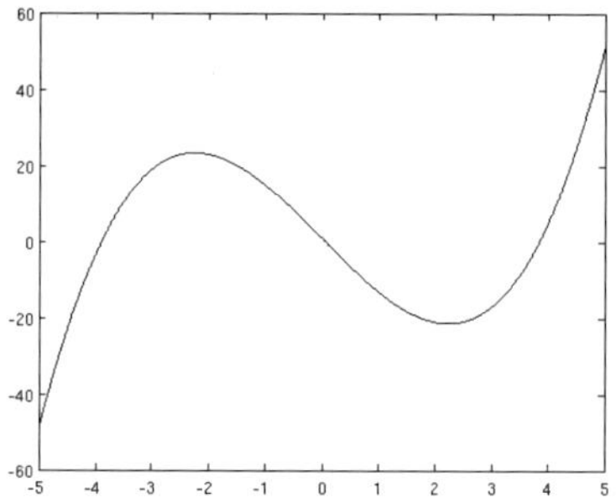

FIGURE 10. The polynomial function χ has 3 distinct roots if χ' has two roots, in which χ takes values on both sides of the axis.

Proof. The proof is analogous to the E-series case, see Lemma 124. □

4.2.2. *Construction of retraction for* $Z_{1,0}(*)$. This construction follows the construction for $E_{2,0}(*)$:

The weighted homogeneous ministable unfolding $F_{(1,0)}\colon \mathbb{R}^{10} \to \mathbb{R}^9$ of $Z_{1,0}(\lambda)$ has one non-positively weighted unfolding variable v_3. The instability locus of $F^+_{(1,0)}\colon \mathbb{R}^9 \to \mathbb{R}^8$ is just the origin by Theorem 156. Define the weighted homogeneous distance function $\rho\colon \mathbb{R}^9 \to \mathbb{R}$ and use the level set restriction $(F_{(1,0)})_\epsilon\colon F^{-1}_{(1,0)}\rho^{-1}(\epsilon) \to \rho^{-1}(\epsilon)$ for some $\epsilon > 0$; now $(F^+_{(1,0)})_\epsilon$ is stable by [dPW95, Lemma 9.6.2]. Hence we can find a smooth retraction

$$(r_\epsilon, s_\epsilon)\colon (F_{(1,0)})_\epsilon \to (F^+_{(1,0)})_\epsilon,$$

and we can extend it to a stratified smooth E-tame retraction

$$(r_{(1,0)}, s_{(1,0)})\colon F_{(1,0)} \to F^+_{(1,0)}$$

using Lemma 36.

The retraction we just constructed is $\mathbb{R}^+$-equivariant, and just as in the $E_{2,0}(*)$ case, we can use the Haar integral with the defining vector fields to make it $\mathbb{R}^*$-equivariant as well.

We have proven:

Theorem 164. *There exists an* $\mathbb{R}^*$*-equivariant, stratified smooth, E-tame retraction* $(r_{(1,0)}, s_{(1,0)})\colon F_{(1,0)} \to F^+_{(1,0)}$. □

Note that the retraction $(r_{(1,0)}, s_{(1,0)})$ and its foliation are not uniquely defined, as we have made a choice as to which smooth retraction to use on the level set.

4.2.3. *Construction of retraction for* $Z_{2,0}(*)$. The weighted homogeneous ministable unfolding $F_{(2,0)}\colon \mathbb{R}^{16} \to \mathbb{R}^{15}$ of $Z_{2,0}(\lambda)$ has two non-positively weighted unfolding variables, v_5 and v_6. By Theorem 156 the instability locus of $F^+_{(2,0)} = F^2_{(2,0)}\colon \mathbb{R}^{14} \to$

$\mathbb{R}^{13}$ is 1-dimensional and has three types of singularities presented in it; namely a $Z_{2,0}(\lambda)$-singularity at the origin and $Z_{1,0}(\lambda).E_{1,0}$- and $E_{2,0}(\lambda).A_1^{c(\lambda)}$- singularities along 1-dimensional, $\mathbb{R}^+$-invariant submanifolds of $\mathbb{R}^{13}$, where

$$c(\lambda) = \begin{cases} 1 & \text{if } \lambda > -\sqrt[3]{\frac{27}{4}}, \\ 3 & \text{if } \lambda < -\sqrt[3]{\frac{27}{4}}. \end{cases}$$

Let ρ be the weighted distance function in $t(F_{(2,0)})$, let $\epsilon > 0$ and denote by $(F_{(2,0)})_\epsilon$ the restriction to the level sets $(F^{-1}_{(2,0)}\rho^{-1}(\epsilon), \rho^{-1}(\epsilon))$. By [dPW95, Lemma 9.6.2], $(F^1_{(2,0)})_\epsilon$ is stable and the instability locus of $(F^2_{(2,0)})_\epsilon$ is $\rho^{-1}(\epsilon) \cap I(F^2_{(2,0)})$, which consists of four points $y_1, \ldots, y_4$ where the $Z_{1,0}(\lambda).E_{1,0}$- and $E_{2,0}(\lambda).A_1^{c(\lambda)}$-singularities are presented, and the stable germ of $(F^1_{(2,0)})_\epsilon$ at y_i is $\mathscr{A}$-equivalent to one of the multigerms

$$\begin{aligned}
&F_{E_{1,0}.Z_{1,0}} := F^E_{(1,0)} \times \mathrm{id}_{t(F^Z_{(1,0)})} \sqcup \mathrm{id}_{t(F^E_{(1,0)})} \times F^Z_{(1,0)} \colon \textstyle\bigsqcup_2 \mathbb{R}^{14} \to \mathbb{R}^{13},\\
&F_{E_{2,0}.A_1^3} := F^E_{(2,0)} \times \mathrm{id}_{\mathbb{R}^3} \sqcup \textstyle\bigsqcup_{i=1}^3 \mathrm{id}_{t(F^E_{(2,0)})} \times \sigma_i \circ (F_{A_1} \times \mathrm{id}_{\mathbb{R}^2}) \colon \textstyle\bigsqcup_4 \mathbb{R}^{14} \to \mathbb{R}^{13},\\
&F_{E_{2,0}.A_1} \times \mathrm{id}_{\mathbb{R}^2} := F^E_{(2,0)} \times \mathrm{id}_{\mathbb{R}} \times \mathrm{id}_{\mathbb{R}^2} \sqcup \mathrm{id}_{t(F^E_{(2,0)})} \times F_{A_1} \times \mathrm{id}_{\mathbb{R}^2} \colon \textstyle\bigsqcup_2 \mathbb{R}^{14} \to \mathbb{R}^{13},
\end{aligned}$$

where $F_{A_1} \colon \mathbb{R}^2 \to \mathbb{R}$ is the stable map $(x, y) \mapsto xy$ and σ_i is a suitable permutation.

We fix coordinates (Ψ, Φ) at y_i such that:

$$\begin{array}{ccc}
\left(s((F^1_{(2,0)})_\epsilon), F^{-1}_{(2,0)}(y_i) \cap \Sigma F_{(2,0)}\right) & \xrightarrow{(F^1_{(2,0)})_\epsilon} & (t((F^1_{(2,0)})_\epsilon), y_i) \\
\Big\downarrow \Psi & & \Big\downarrow \Phi \\
s(F_\Delta) & \xrightarrow[F_\Delta]{} & t(F_\Delta)
\end{array}$$

for $\Delta \in \{E_{1,0}.Z_{1,0}(\lambda), E_{2,0}(\lambda).A_1^{c(\lambda)}\}$. Here F_Δ is the standard ministable unfolding of the multigerm Δ as usual, except for the case where $\Delta = E_{2,0}(\lambda).A_1$. In this case, we replace F_Δ with its suspension $F_{E_{2,0}.A_1} \times \mathrm{id}_{\mathbb{R}^2}$.

Using Lemma 29 and the previously constructed E-tame retractions for Z- and E-series singularities, we construct an E-tame retraction

$$(r_\Delta, s_\Delta) \colon F_\Delta \to F^+_\Delta.$$

The fibers of s_Δ are transverse to $t\left((F^+_{(3,0)})_\epsilon\right)$ in $t\left((F_{(3,0)})_\epsilon\right)$ by Proposition 163, since the fibers are ST-invariant. For the $E_{2,0}(\lambda).A_1$-case we note, in addition, that the fiber $s_\Delta^{-1}(y)$ is transverse to the presentation

$$(A_1.E_{2,0}(\lambda))_{\text{strict}}\,(F^+)$$

in the presentation

$$(A_1.E_{2,0}(\lambda))_{\text{strict}}\,(F).$$

Thus, by Proposition 26, this retraction induces an E-tame retraction

$$(F_{(2,0)})_\epsilon \to (F^+_{(2,0)})_\epsilon$$

near y_i, which has the same fibers as (r_Δ, s_Δ).

On the other hand, away from y_i we can find smooth retractions

$$(F_{(2,0)})_{\hat\epsilon} \to (F^+_{(2,0)})_{\hat\epsilon}.$$

Using Lemma 32 we may combine the local retractions to obtain an E-tame retraction $(F_{(2,0)})_\epsilon \to (F^+_{(2,0)})_\epsilon$ and by Lemma 36 we obtain an E-tame retraction

$$(r_{(2,0)}, s_{(2,0)})\colon F_{(2,0)} \longrightarrow F^+_{(2,0)}.$$

We have proven:

Theorem 165. *We can construct an E-tame retraction*

$$(r_{(2,0)}, s_{(2,0)})\colon F_{(2,0)} \longrightarrow F^+_{(2,0)}.$$

□

Note again that, in spite of our notation, our method does not define the retraction $(r_{(2,0)}, s_{(2,0)})$ uniquely, as we have made choices regarding which smooth retractions to use in the very beginning.

4.2.4. *Construction of retraction for $Z_{3,0}(*)$.* The weighted homogeneous ministable unfolding $F_{(3,0)}\colon \mathbb{R}^{22} \to \mathbb{R}^{21}$ has three non-positively weighted unfolding variables v_7, v_8, v_9, and, as in the $E_{4,0}(*)$ case, the positive instability locus is a stratified set; we shall analyze it more carefully in a second. Just like we did in the $E_{4,0}(*)$ case, we will construct the retraction $F_{(3,0)} \to F^+_{(3,0)}$ by first restricting to a level set of the weighted homogeneous distance function $\rho_{(3,0)}$. Here we find presented combinations of $E_{q,0}(\lambda)$-singularities with $q \leq 3$, and $Z_{q,0}(\lambda)$-singularities, all with $q \leq 2$, suggesting an inductive construction of the retraction. We can find local E-tame retractions using the previous results, but in order to combine them we need – as in the $E_{4,0}(*)$ case – to control the geometry near the instability locus.

Pass to a slice

Let $\rho_{(3,0)}\colon \mathbb{R}^{21} \to \mathbb{R}$ be the weighted distance function as defined in Chapter 2.4, pick $\epsilon > 0$, and restrict to the level sets $(F^{-1}_{(3,0)}\rho^{-1}_{(3,0)}(\epsilon), \rho^{-1}_{(3,0)}(\epsilon))$. We denote the restricted map by $(F_{(3,0)})_\epsilon$; similarly we denote by $(F^+_{(3,0)})_\epsilon$ the restriction of $F^+_{(3,0)}$ to

$$\left(F^{-1}_{(3,0)}\rho^{-1}_{(3,0)}(\epsilon) \cap s(F^+_{(3,0)}), \rho^{-1}_{(3,0)}(\epsilon) \cap t(F^+_{(3,0)})\right).$$

If we can find an E-tame retraction $(r, s)\colon (F_{(3,0)})_\epsilon \to (F^+_{(3,0)})_\epsilon$, then we can find an E-tame retraction $(R, S)\colon F_{(3,0)} \to F^+_{(3,0)}$ by Lemma 36.

By [dPW95, Lemma 9.6.2], $(F^1_{(3,0)})_\epsilon$ is stable, and we can find a smooth retraction $(F_{(3,0)})_\epsilon \to (F^1_{(3,0)})_\epsilon$. By [dPW95, Lemma 9.3.22], it suffices to find an E-tame retraction $(F^1_{(3,0)})_\epsilon \to (F^+_{(3,0)})_\epsilon$ in order to get an E-tame retraction $(F_{(3,0)})_\epsilon \to (F^+_{(3,0)})_\epsilon$.

For simplicity, we agree to denote $(F^1_{(3,0)})_\epsilon$ by F, and $(F^+_{(3,0)})_\epsilon$ by F^+ through the whole construction.

The positive instability locus

As discussed earlier, the instability locus of $F^+_{(3,0)}$ is parametrized by the (x,y)-polynomial coefficients of the deformation $\mathbb{R}^p \to \mathbb{R}$ given by

$$x\left(y^3 + \lambda y(x-\xi_1)^2(x-\xi_2)^2(x-\xi_3)^2 + (x-\xi_1)^3(x-\xi_2)^3(x-\xi_3)^3\right)$$

on subsets of $\{(\xi_1,\xi_2,\xi_3) | \xi_1 \le \xi_2 \le \xi_3\}$ where $\{\xi_i = 0\}$, $\{\xi_1 = \xi_2\}$ or $\{\xi_2 = \xi_3\}$.

Thus, we get components of strata of the positive instability locus defined by subsets of parameter space with the presented singularity types displayed in the table below, and we give a geometric sketch of its intersection with $\rho^{-1}_{(3,0)}(\epsilon)$ in Figure 11:

	stratum in $\mathbb{R}^p$	**dimension**	**singularity type**
	$\{\xi_1 = \xi_2 = \xi_3\}$	0 (origin)	$Z_{3,0}(\lambda)$
A)	$\{\xi_1 = \xi_2 = 0 < \xi_3\}$	1	$Z_{2,0}(\lambda).E_{1,0}$
B)	$\{\xi_1 = 0 < \xi_2 = \xi_3\}$	1	$Z_{1,0}(\lambda).E_{2,0}(\lambda)$
C)	$\{\xi_1 = \xi_2 = \xi_3 > 0\}$	1	$A_1^{c(\lambda)}.E_{3,0}(\lambda)$
D)	$\{\xi_1 < \xi_2 = \xi_3 = 0\}$	1	$Z_{2,0}(\lambda).E_{1,0}$
E)	$\{\xi_1 = \xi_2 < \xi_3 = 0\}$	1	$Z_{1,0}(\lambda).E_{2,0}(\lambda)$
F)	$\{\xi_1 = \xi_2 = \xi_3 < 0\}$	1	$A_1^{c(\lambda)}.E_{3,0}(\lambda)$
1)	$\{\xi_1 = 0 < \xi_2 < \xi_3\}$	2	$Z_{1,0}(\lambda).E_{1,0}^2$
2)	$\{0 < \xi_1 = \xi_2 < \xi_3\}$	2	$A_1^{c(\lambda)}.E_{2,0}(\lambda).E_{1,0}$
3)	$\{\xi_1 < \xi_2 = \xi_3 < 0\}$	2	$A_1^{c(\lambda)}.E_{2,0}(\lambda).E_{1,0}$
4)	$\{\xi_1 < \xi_2 = 0 < \xi_3\}$	2	$Z_{1,0}(\lambda).E_{1,0}^2$
5)	$\{\xi_1 = \xi_2 < \xi_3 = 0\}$	2	$A_1^{c(\lambda)}.E_{2,0}(\lambda).E_{1,0}$
6)	$\{\xi_1 < 0 < \xi_2 = \xi_3\}$	2	$A_1^{c(\lambda)}.E_{2,0}(\lambda).E_{1,0}$
7)	$\{\xi_1 < \xi_2 < \xi_3 < 0\}$	2	$Z_{1,0}(\lambda).E_{1,0}^2$
8)	$\{\xi_1 = \xi_2 < 0 < \xi_3\}$	2	$A_1^{c(\lambda)}.E_{2,0}(\lambda).E_{1,0}$
9)	$\{0 < \xi_1 < \xi_2 = \xi_3\}$	2	$A_1^{c(\lambda)}.E_{2,0}(\lambda).E_{1,0}$

where

$$c(\lambda) = \begin{cases} 1 & \text{if } \lambda > -\sqrt[3]{\frac{27}{4}} \\ 3 & \text{if } \lambda < -\sqrt[3]{\frac{27}{4}} \end{cases}$$

From now on we work inside the level set $\rho^{-1}_{(3,0)}(\epsilon)$, which is transverse to all these strata – hence the dimensions will be one lower than those displayed in the table.

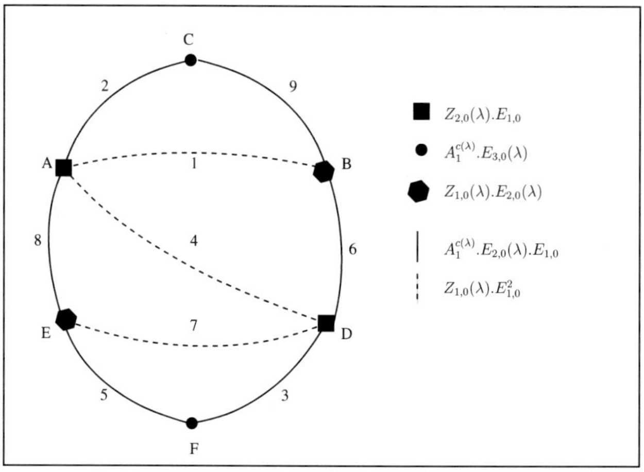

FIGURE 11. The stratified positive instability locus of $F_{(3,0)}$, intersected with the level set $\rho_{(3,0)}^{-1}(\epsilon)$.

We note that the 1-dimensional strata (corresponding to the 2-dimensional strata in the table above) are deformations of *one* of the singularities appearing in 0-dimensional strata (1-dimensional in the table). The 0-dimensional strata are multi-germs where at most two are deformable within the positive instability locus; in the deformations giving rise to 1-dimensional strata, one is kept constant while the other is deformed.

We shall use $\tilde{\Delta} = \prod_{i=1}^{\tilde{s}} \tilde{\Delta}_i$ to denote the singularity types presented on 1-dimensional strata, and $\Delta = \prod_{i=1}^{s} \Delta_i$ to denote the singularity types presented on 0-dimensional strata.

We denote the strata of the instability locus by I_Δ and $I_{\tilde{\Delta}}$, where Δ and $\tilde{\Delta}$ denote the presented singularities, and we denote the corresponding subsets of source by H_Δ and $H_{\tilde{\Delta}}$. The full presentations in the target of F are denoted by Y_Δ and $Y_{\tilde{\Delta}}$.

Choice of coordinates at I_Δ

Consider a 0-dimensional stratum I_Δ, with $y \in I_\Delta$, and fix coordinates $(\psi_\Delta, \phi_\Delta)$ such that

$$\begin{array}{ccc} (s(F), F^{-1}(y) \cap \Sigma F) & \xrightarrow{F} & (t(F), y) \\ \Big\downarrow{\scriptstyle \psi_\Delta} & & \Big\downarrow{\scriptstyle \phi_\Delta} \\ (s(F_\Delta), \bigsqcup 0) & \xrightarrow[F_\Delta]{} & t(F_\Delta) \end{array}$$

where F_Δ is the standard ministable multigerm unfolding of the singularity Δ, as defined in Lemma 11, unless $\Delta = A_1.E_{3,0}(\lambda)$, in which case we replace F_Δ with its suspension $F_{A_1.E_{(3,0)}} \times \mathrm{id}_{\mathbb{R}^2}$. Note that we can write $\Delta = \Delta_1.\Delta_2$ where either Δ_i is a monogerm or $\Delta_i = A_1^{c(\lambda)}$, and where the "incoming" 1-dimensional strata are deformations of one of the Δ_i. Note furthermore that $t(F_\Delta) = t(F_{\Delta_1}) \times t(F_{\Delta_2})$, and we can define the distance function

$$\rho_{\Delta,\tilde{\Delta}} = \rho_{\Delta_i} \circ \mathrm{pr}_{t(F_{\Delta_i})},$$

where Δ_i is the singularity that deforms to create the stratum $I_{\tilde{\Delta}}$.

We denote by $F^1_{\Delta,\tilde{\Delta}}$ the germ obtained from F_Δ by removing the most negatively weighted unfolding variable from F_{Δ_i}, where Δ_i is the singularity which deforms to create $\tilde{\Delta}$. This is analogous to what we did for $E_{4,0}(\lambda)$.

The level sets of these distance functions define local tubular neighborhoods $(T_{\tilde{\Delta},\Delta}, \pi_{\tilde{\Delta},\Delta})$ of $I_{\tilde{\Delta}}$ near I_Δ, and they pull back to source as in the $E_{4,0}(\lambda)$-case.

Find suitable tubular neighborhoods of the 1-dimensional strata

Using similar techniques as in Section 4.1.6, we can find tubular neighborhoods $(T_{\tilde{\Delta}}, \pi_{\tilde{\Delta}})$ of $I_{\tilde{\Delta}}$, such that

- The tubular neighborhoods $(T_{\tilde{\Delta}}, \pi_{\tilde{\Delta}})$ pull back to tubular neighborhoods $(\tilde{T}_{\tilde{\Delta}}, \tilde{\pi}_{\tilde{\Delta}})$ of $H_{\tilde{\Delta}}$ in source.
- Near I_Δ, the tubular neighborhoods
  $$\left((\tilde{T}_{\tilde{\Delta}}, \tilde{\pi}_{\tilde{\Delta}}), (T_{\tilde{\Delta}}, \pi_{\tilde{\Delta}})\right)$$
  coincide with the tubular neighborhood
  $$\left((\tilde{T}_{\tilde{\Delta},\Delta}, \tilde{\pi}_{\tilde{\Delta},\Delta}), (T_{\tilde{\Delta},\Delta}, \pi_{\tilde{\Delta},\Delta})\right)$$
  defined by the level sets of the distance function $\rho_{\Delta,\tilde{\Delta}}$.
- The map $F|\tilde{T}_{\tilde{\Delta}}$ is trivial over $I_{\tilde{\Delta}}$.
- The restriction of F to any fiber of $\left(\tilde{T}_{\tilde{\Delta}}, \tilde{\pi}_{\tilde{\Delta}}\right)$ is stable.

We can find submanifolds $M_{\tilde{\Delta}}$ of $T_{\tilde{\Delta}}$ which pull back to submanifolds $\tilde{M}_{\tilde{\Delta}}$ of $\tilde{T}_{\tilde{\Delta}}$, both of codimension $d(\tilde{\Delta}, \lambda)$, where

$$d(\tilde{\Delta}, \lambda) = \begin{cases} 1 & \text{if } \tilde{\Delta} = Z_{1,0}(\lambda).E^2_{1,0} \text{ or } \lambda < -\sqrt[3]{\frac{27}{4}} \\ 3 & \text{if } \tilde{\Delta} = A_1.E_{2,0}(\lambda).E_{1,0} \text{ and } \lambda > -\sqrt[3]{\frac{27}{4}} \end{cases}$$

such that

- Near the 0-dimensional strata I_Δ,
  $$\left(\psi_\Delta(\tilde{M}_{\tilde{\Delta}}), \phi_\Delta(M_{\tilde{\Delta}})\right) = \left(s(\tilde{F}^1_{\tilde{\Delta}}), t(\tilde{F}^1_{\tilde{\Delta}})\right),$$
  where $\tilde{F}$ denotes the ministable unfolding of Δ in *all* cases.

- The restriction $F|\tilde{M}_{\tilde{\Delta}}$ is stable.
- The tubular neighborhood structure from $(\tilde{T}_{\tilde{\Delta}}, T_{\tilde{\Delta}})$ restricts to a tubular neighborhood structure on $(\tilde{M}_{\tilde{\Delta}}, M_{\tilde{\Delta}})$ for all $\tilde{\Delta}$.
- The restriction $F_M = F|\colon \tilde{M}_{\tilde{\Delta}} \to M_{\tilde{\Delta}}$ is stable for each $\tilde{\Delta}$.
- The map F_M is trivial over $I_{\tilde{\Delta}}$, and the restriction to any fiber at $I_{\tilde{\Delta}}$ is ministable and equivalent to $F_{\tilde{\Delta}}$ for all $\tilde{\Delta}$.

Find retractions at the 0-dimensional strata

We pull back the retractions (r_Δ, s_Δ) to a neighborhood of I_Δ using $(\psi_\Delta, \phi_\Delta)$. Note that if $y \in I_{\tilde{\Delta}} \cap \rho^{-1}_{\Delta,\tilde{\Delta}}(\epsilon)$ for a sufficiently small ϵ, then the restriction of (r_Δ, s_Δ) to the subset

$$\Big((F_\Delta^{-1}\rho^{-1}_{\Delta,\tilde{\Delta}}(\epsilon), F_\Delta^{-1}(y) \cap \Sigma F_\Delta), (\rho^{-1}_{\Delta,\tilde{\Delta}}(\epsilon), y)\Big) \bigcap \big(s(F^1_\Delta), t(F^1_\Delta)\big)$$

is taken to $(s_{\tilde{\Delta}}, r_{\tilde{\Delta}})$ by $(\psi_\Delta, \phi_\Delta)$, by the construction of the retractions (r_Δ, s_Δ). This is analogous to the $E_{4,0}(\lambda)$ case.

Find retractions near the positive instability locus

As in the E-series case, we now have local E-tame retractions (r_Δ, s_Δ) at the 0-dimensional strata (H_Δ, I_Δ), which restrict to E-tame retractions in the fibers of the tubular neighborhoods on the submanifolds $(\tilde{M}_{\tilde{\Delta},\Delta}, M_{\tilde{\Delta},\Delta})$ of the 1-dimensional strata $(H_{\tilde{\Delta}}, I_{\tilde{\Delta}})$ near the (H_Δ, I_Δ). We also have a different choice of fibered E-tame retractions in the global submanifolds $(\tilde{M}_{\tilde{\Delta}}, M_{\tilde{\Delta}})$ of the $(H_{\tilde{\Delta}}, I_{\tilde{\Delta}})$ – hence we get two different choices of E-tame retractions near the (H_Δ, I_Δ), stemming from two different choices of weighted homogeneous coordinates in the fiber of the tubular neighborhood.

As in the E-series case, we can interpolate between the choices of coordinates along the $I_{\tilde{\Delta}}$ using the contractibility of $\mathscr{A}_{F_{\tilde{\Delta}}}/MC(\mathscr{A}_{F_{\tilde{\Delta}}})$. This gives us a new choice of coordinates in the fibers of $(\tilde{M}_{\tilde{\Delta}}, M_{\tilde{\Delta}})$. Near I_Δ, this gives the same coordinates, and the same retraction, as that defined by the (r_Δ, s_Δ), modulo an element of a maximal compact subgroup G of $\mathscr{A}_{F_{\tilde{\Delta}}}$. By Theorem 86 we can choose this group to be a product of subgroups $MC(\mathscr{K}_{f_{\tilde{\Delta}_i}})$, one for each monogerm appearing in $\tilde{\Delta}$.

Recall that by Theorem 97, $MC(\mathscr{A}_{F_{\tilde{\Delta}_i}}) = \{\pm 1\}$ for $\tilde{\Delta}_i = E_{p,0}(\lambda)$ or $Z_{p,0}(\lambda)$. The multigerm $\tilde{\Delta}$ consists of singularities $E_{1,0}$, $E_{2,0}(\lambda)$, $Z_{1,0}(\lambda)$, A_1^3; the only ones whose retractions are not identities, are the $E_{2,0}(\lambda)$ and $Z_{1,0}(\lambda)$. Both of these admit $\{\pm 1\}$-equivariant E-tame retractions by Theorems 133 and 164. Then $(r_{\tilde{\Delta}}, s_{\tilde{\Delta}})$ is a G-equivariant, E-tame retraction. In particular, its fibers coincide with those of the retraction defined by (r_Δ, s_Δ) in $(\tilde{M}_{\tilde{\Delta}}, M_{\tilde{\Delta}})$ near (H_Δ, I_Δ). These retractions combine with smooth projections from $(\tilde{T}_{\tilde{\Delta}}, T_{\tilde{\Delta}})$ onto $(\tilde{M}_{\tilde{\Delta}}, M_{\tilde{\Delta}})$ to create E-tame retractions which coincide with the retractions induced by (r_Δ, s_Δ) near the (H_Δ, I_Δ).

Moreover, as in the E-series case we can show that its fibers are transverse to $t((F^+_{(3,0)})_\epsilon)$ in $t((F_{(3,0)})_\epsilon)$, and hence by Proposition 26, we get a local retraction $(F_{(3,0)})_\epsilon \to (F^+_{(3,0)})_\epsilon$.

It follows that we can construct an E-tame retraction $(F_{(3,0)})_\epsilon \to (F^+_{(3,0)})_\epsilon$ near the instability locus, and – as in the E-series case – we can combine it with smooth retractions off the instability locus, in order to obtain an E-tame retraction $(F_{(3,0)})_\epsilon \to (F^+_{(3,0)})_\epsilon$. This finishes the construction.

We have proven:

Theorem 166. *There exists an E-tame retraction*

$$(r_{(3,0)}, s_{(3,0)}) \colon F_{(3,0)} \to F^+_{(3,0)},$$

and $F^+_{(3,0)}$ is topologically ministable. □

4.3. **E-tame retractions for the non-weighted homogeneous cases.** Our E-tame retractions $F_{(p,0)} \to F^+_{(p,0)}$ in the E- and Z-series have all been constructed inductively by extending retractions on a level set using Lemma 36. By this lemma, we find E-tame retractions and corresponding foliations $(\mathscr{F}_s, \mathscr{F}_t)$ in $(s(F_{(p,0)}), t(F_{(p,0)}))$ in neighborhoods (W_M, W_N) of the subspaces of source and target that have positive weights.

Furthermore, the germs from (105) and (153) with $w_0 \neq 0$ and (153) all appear as germs of $F_{(p,0)}$ for suitable $w_0 \neq 0$ at points $(0,0,u) \in t(F^+_{(p,0,0)}) \times U_0 \times U_-$ on the respective negatively weighted subspaces. By restricting to the subspaces

$$(s(F^+_{(p,0)}) \times \{0\} \times \{u\}, t(F^+_{(p,0)}) \times \{0\} \times \{u\}) \subset \left(s(F_{(p,0)}), t(F_{(p,0)})\right),$$

we get the wanted germs from (105) and (153). The non-positively weighted subspaces in $s(F_{(p,0)})$ and $t(F_{(p,0)})$ are leaves of $\mathscr{F}_s$ and $\mathscr{F}_t$, respectively, thus the germs at $((0,u),(0,u))$ of the foliations $(\mathscr{F}_s, \mathscr{F}_t)$ are transverse to

$$\left(s(F^+_{(p,0)}) \times \{0\} \times \{u\}, t(F^+_{(p,0)}) \times \{0\} \times \{u\}\right),$$

and they define E-tame retractions

$$F \to F| \left((s(F^+_{(p,0)}) \times \{0\} \times \{u\}, (0,0,u)), (t(F^+_{(p,0)}) \times \{0\} \times \{u\}, (0,0,u))\right).$$

But these are just E-tame retractions onto the positively weighted part of the standard ministable unfolding for (105) and (153).

4.4. **Conclusion and future work.** We conjecture that it is possible to construct E-tame retractions for $E_{p,0}(*)$ and $Z_{q,0}(*)$ for all $p, q \in \mathbb{N}$, having seen that it *is* possible for $p \leq 4$ and $q \leq 3$.

The constructions of the E-tame retractions follow the plan

Step 1 Parametrize the positive instability locus, and show that it is a stratified set with respect to stratification by presented singularity type. Identify the presented singularity types in the positive instability locus, and realize that by induction, we already know how to find E-tame retractions onto their positively weighted unfoldings.

Step 2 Use this to find local retractions near all points of the instability locus.

Step 3 Combine the local retractions by controlling the geometry near the positive instability locus, hence forcing the local retractions to coincide on common domains.

We have already completed Step 1 for all p and q. In Step 3 we suggest constructing a system of trivial tubular neighborhoods about strata in the spirit of the $E_{4,0}(\lambda)$ case, and using the contractibility results found in Chapter 3 to choose suitable coordinates in the tubular neighborhood fibers.

Step 2 suggests an inductive procedure, but as we have seen in the second construction for the $E_{4,0}(*)$ and $Z_{3,0}(*)$ cases, we need our retractions $F_{(p,0)} \to F^+_{(p,0)}$ to be equivariant with respect to $MC(\mathscr{A}_{F_{(p,0)}})$ for $F_{(p,0)}$ occurring on dim > 1 strata of the positive instability locus, in order to force retractions to coincide using the contractibility of quotients. Thus, in attacking the general problem, we need to make sure that we can construct *equivariant* retractions in each step.

INDEX

References

[BCFR07] Birbrair, L., J. C. F. Costa, A. Fernandes and M. A. S. Ruas: $\mathscr{K}$-bi-Lipschitz equivalence of real function-germs. - Proc. Amer. Math. Soc. 135, no. 4, 2007, 1089–1095.

[Boc45] Bochner, S.: Compact groups of differentiable transformations. - Ann. of Math. 46, no. 3 1945, 372–381.

[CL55] Coddington, E. A. and N. Levinson: Theory of ordinary differential equations. - McGraw-Hill, inc., New York-Toronto-London, 1955.

[Dam88] Damon, J. N.: Topological triviality and versality for subgroups of $\mathscr{A}$ and $\mathscr{K}$. - Mem. Amer. Math. Soc. 389, 1988.

[Dam92] Damon, J. N.: Topological triviality and versality for subgroups of $\mathscr{A}$ and $\mathscr{K}$. II. Sufficient conditions and applications. - Nonlinearity 5, no. 2, 1992, 373–412.

[Dam95] Damon, J. N.: Local Morse theory for solutions to the heat equation and Gaussian blurring. - J. Differential Equations, 115, no. 2, 1995, 368–401.

[Dam99] Damon, J. N.: Generic structure of two-dimensional images under Gaussian blurring. - SIAM J. Appl. Math., 59, no. 1, 1999, 97–138.

[DG93] Damon, J. N. and A. Galligo: Universal topological stratification for the Pham example. - Bull. Soc. Math. France, 121, no. 2, 1993, 153–181.

[DG99] Donelan, P. S. and C. G. Gibson: Singular phenomena in kinematics. - Singularity theory (Liverpool, 1996), London Math. Soc. Lecture Note Ser. 263, Cambridge Univ. Press, 1999, 379–402.

[dP] du Plessis, A. A.: Genericity and smooth finite determinacy. - Singularities, Part 1 (Arcata, Calif., 1981), Proc. Sympos. Pure Math. 40, 295–312.

[dP99] du Plessis, A. A.: Finiteness of Mather's canonical stratification. - Singularity theory (Liverpool, 1996), London Math. Soc. Lecture Note Ser. 263, 1999, 199–206.

[dPW] du Plessis, A. A. and L. C. Wilson: Right-symmetry of mappings. - Singularity theory and its applications, Part I, Lecture Notes in Math. 1462, Springer, 1988/1989, 258–275.

[dPW95] du Plessis, A. A. and C. T. C. Wall: The geometry of topological stability. - London Mathematical Society Monographs New Series. 9, 1995.

[dPW04] du Plessis, A. A. and C. T. C. Wall: Topology of unfoldings of singularities in the E, Z and Q series. - Real and complex singularities, Contemp. Math. 354, 2004, 227–258.

[FR04] Fernandes, A. and M. A. S. Ruas: Bi-Lipschitz determinacy of quasihomogeneous germs. - Glasg. Math. J., 46, no. 1, 2004, 77–82.

[GG73] Golubitsky, M. and V. Guillemin: Stable mappings and their singularities. - Graduate Texts in Mathematics. 14, Springer, 1973.

[GWdPL76] Gibson, C. G., K. Wirthmüller, A. A. du Plessis and E. J. N. Looijenga: Topological stability of smooth mappings. - Lecture Notes in Mathematics. 552, Springer, 1976.

[Hir76] Hirsch, M. W.: Differential topology. - Graduate Texts in Mathematics. 33, Springer, 1976.

[HMdJRF07] Hacon, D., C. Mendes de Jesus and M. C. Romero Fuster: Stable maps from surfaces to the plane with prescribed branching data. - Topology Appl. 154, no. 1, 2007, 166–175.

[Hoc65] Hochschild, G.: The structure of Lie groups. - Holden-Day Inc, San Francisco, 1965.

[Jän78] Jänich, K.: Symmetry properties of singularities of C^∞-functions. - Math. Ann., 238, no. 2, 1978, 147–156.

[Lan95] LANG, S.: Differential and Riemannian manifolds. - Graduate Texts in Mathematics, Vol. 160, Springer, 3. ed., 1995.

[Lee03] LEE, J. M.: Introduction to smooth manifolds. - Graduate Texts in Mathematics. Vol. 218, Springer, 2003.

[Loo77] LOOIJENGA, E.: On the semi-universal deformation of a simple-elliptic hypersurface singularity. Unimodularity. - Topology 16, no. 3, 1977, 257–262.

[Mat68b] MATHER. J. N.: Stability of C^∞ mappings. I. The division theorem. - Ann. of Math. no. 2, 87, 1968, 89–104.

[Mat69b] MATHER. J. N.: Stability of C^∞ mappings. II. Infinitesimal stability implies stability. - Ann. of Math. no. 2, 89, 1969, 254–291.

[Mat68a] MATHER. J. N.: Stability of C^∞ mappings. III. Finitely determined mapgerms. - Inst. Hautes Études Sci. Publ. Math., no. 35, 1968, 279–308.

[Mat69a] MATHER. J. N.: Stability of C^∞ mappings. IV. Classification of stable germs by R-algebras. - Inst. Hautes Études Sci. Publ. Math., no. 37, 1969, 223–248.

[Mat70b] MATHER. J. N.: Stability of C^∞ mappings. V. Transversality. - Advances in Math. 4, 1970, 301–336.

[Mat71] MATHER. J. N.: Stability of C^∞ mappings. VI: The nice dimensions. - Proceedings of Liverpool Singularities-Symposium, I (1969/70), Lecture Notes in Math. 192, 1971, 207–253.

[Mat70a] MATHER. J. N.: Notes on topological stability. - Unpublished notes, Harvard, 1970.

[MT06] MUROLO, C. and D. J. A. TROTMAN: Semidifférentiabilité et version lisse de la conjecture de fibration de Whitney. - Singularity theory and its applications, Adv. Stud. Pure Math. 43, 2006, 271–309.

[ON06] OLSEN, O. F. and M. NIELSEN: The generic structure of the optic flow field. - J. Math. Imaging Vision 24, no. 1, 2006, 37–53.

[ORF09] OSET SINHA, R. and M. C. ROMERO FUSTER: Graphs of stable maps from 3-manifolds to $\mathbb{R}^3$. - Preprint, 2009.

[ORF09'] OSET SINHA, R. and M. C. ROMERO FUSTER: Global topological invariants of stable maps from 3-manifolds to $\mathbb{R}^3$. - Proceedings of the Steklov Institute of Mathematics, to appear.

[Rim96] RIMÁNYI, R.: Generalized Pontrjagin-Thom construction for singular maps. - Thesis, Eötvös Loránd University, Budapest, 1996.

[Rim02] RIMÁNYI, R.: On right-left symmetries of stable singularities. - Math. Z., 242, no. 2, 2002, 347–366.

[Sch89] SCHWARZ, G. W.: The topology of algebraic quotients. - Topological methods in algebraic transformation groups, Progr. Math. 80, 1989, 135–151.

[Sku09] SKUTLABERG, O.: Topics in topological equivalence of plane-to-plane map-germs; sufficiency of jets and classification. - PhD Thesis, University of Oslo, 2009.

[Slo78] SLODOWY, P.: Einige Bemerkungen zur Entfaltung symmetrischer Funktionen. - Math. Z., 158, no. 2, 1978, 157–170.

[Tou71] TOUGERON, J.-C.: Idéaux de fonctions différentiables. - Actes du Congrès International des Mathématiciens (Nice, 1970), Tome 2, 1971, 267–271.

[Wal80] WALL, C. T. C.: A second note on symmetry of singularities. - Bull. London Math. Soc. 12, no. 5, 1980, 347–354.

[Wal81] WALL, C. T. C.: Finite determinacy of smooth map-germs. - Bull. London Math. Soc. 13, no. 6, 1981, 481–539.

P.O. BOX 68 (GUSTAF HÄLLSTRÖMIN KATU 2B), FI-00014 UNIVERSITY OF HELSINKI, FINLAND

E-mail address: aasa.feragen@helsinki.fi